Key Stage 3 ICT

Control and **Monitoring** with **Flowol2**

P Evans

Published by

PAYNE-GALLWAY
PUBLISHERS LTD

26-28 Northgate Street, Ipswich IP1 3DB
Tel: 01473 251097 • Fax: 01473 232758
www.payne-gallway.co.uk

A catalogue entry for this book is available from the British Library.

ISBN 1 903112 53 2

Published by Payne-Gallway Publishers Limited
26-28 Northgate Street, Ipswich IP1 3DB

Tel: +44 (0)1473 251097 • Fax: +44 (0)1473 232758 • E-mail: info@payne-gallway.co.uk

Reprinted 2004

10 9 8 7 6 5 4 3 2

Design & Artwork by Direction Advertising & Design Ltd

Printed in Great Britain by MWL Print Group Ltd, Pontypool, South Wales

Contents

Preface

This book is designed to introduce pupils to the fundamentals of computer control using Flowol2, which provides a cost-effective way of teaching control in a simulated on-screen environment without the need for additional hardware.

The book is based on the **'Analysing and automating processes'** and **'Control and monitoring'** objectives for Year 7 found within the **'Framework for teaching ICT capability: Years 7, 8** and 9'. The text links with **Sample Teaching Unit 7.6 for ICT**, some aspects of which schools may have adopted within their Key Stage 3 scheme of work. Further information on how this text links with Sample Teaching Unit 7.6 can be downloaded from www.payne-gallway.co.uk/flowol

The book is designed to allow pupils to perform tasks independently within the context of a structured ICT lesson. Additional and extension tasks are included throughout the text to provide opportunities for progression to higher levels of achievement. Model solutions to tasks can be downloaded from the teacher's section at www.payne-gallway.co.uk/flowol

Pupils will need access to computers with Flowol2 and the secondary mimics installed to work through the exercises in this book. You will also need to install the additional mimics fridge, house1 and house2 specifically developed for use with this book. These mimics can be downloaded along with installation instructions from the resources section at www.payne-gallway.co.uk/flowol

The pupil resources referred to throughout the text are freely available from the resources section at www.payne-gallway.co.uk/flowol
Further information on Flowol2 can be found on the following websites:
http://www.data-harvest.co.uk/control/secondary.html
http://www.flowol.com
Flowol2 screenshots are reproduced by kind permission of Rod Bowker - Keep I.T. Easy (K.I.T.E).

Chapter 1 Flowcharts and Problem Solving

In this chapter you will learn how to design and solve computer control problems by breaking them down into simpler steps using diagrams called **flowcharts**.

Before you get started make sure you have a copy of **Worksheet 1**.

Computers are used to control many devices both at home and in the world around us, from washing machines to the autopilot on the aeroplane that takes you on holiday. Anything that can be turned on or off can be controlled by a computer. Lights and buzzers are examples of some of the more simple devices that you might have used already to learn about control at primary school.

Flowcharts

Every computer-controlled system needs a sequence of instructions called a **program**. Programs must have clear instructions in the correct order that tell the computer exactly what to do and when to do it. It is important to plan the instructions in a program carefully so what should happen does happen. **Flowcharts** are diagrams that use special symbols to describe the steps needed in a program and the order in which they must be carried out. Breaking problems into simpler steps and representing them in diagrams like flowcharts helps us to plan the most efficient solutions by showing clearly what must happen.

Flowchart symbols

The symbols used in flowcharts are described below.

START or STOP

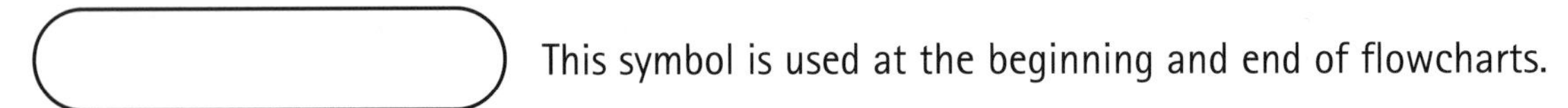

This symbol is used at the beginning and end of flowcharts.

The symbols in flowcharts are connected using arrows. START and STOP symbols may have only one connecting arrow going in to or out of them, as shown in the example below about watering plants.

PROCESS

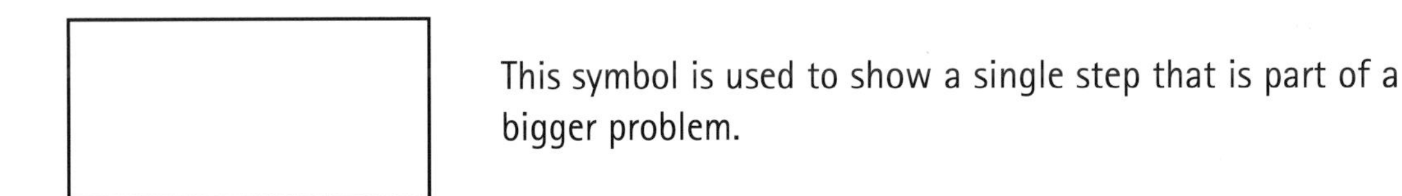

This symbol is used to show a single step that is part of a bigger problem.

PROCESS symbols may have only one line going into and one line going out of them.

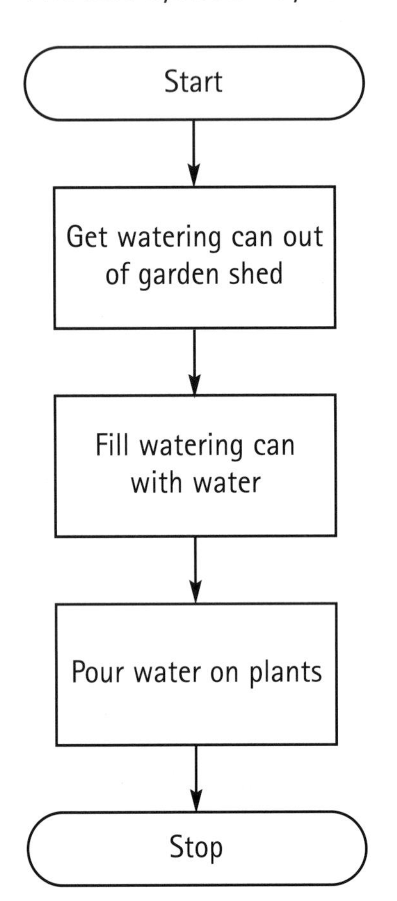

In this example the problem of how to water plants in a garden has been broken down into three steps. These steps are represented in the flowchart below by three PROCESS symbols.

DECISION

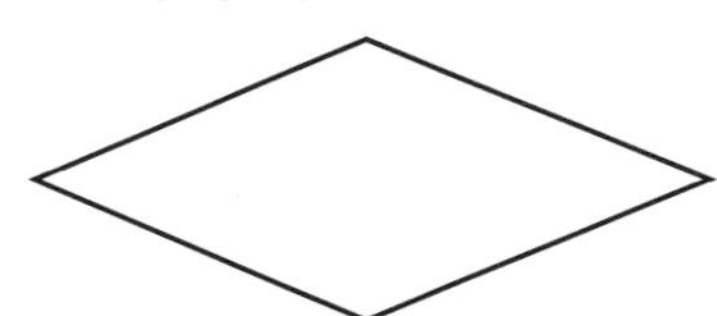

This symbol is used in a flowchart when a question must to be asked to decide what to do next.

DECISION symbols must have one YES Arrow and one NO Arrow going out of them to show what must happen next depending on the answer to the question. Here is an example. This flowchart describes the problem of how to cross the road.

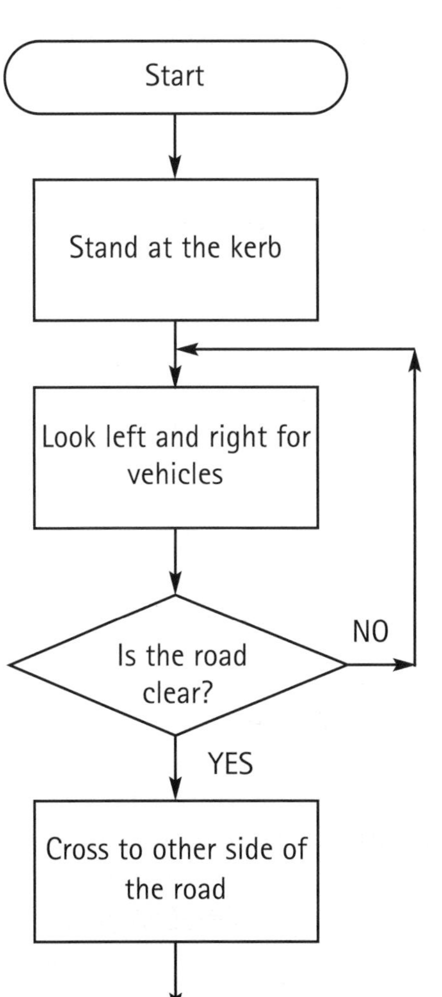

A DECISION symbol is used to ask a question that checks there are no vehicles coming before stepping out into the road.

The DECISION symbol in this flowchart has two arrows coming out of it labelled YES and NO.

If the answer to the question is NO, arrows lead back to a place in the flowchart just before the PROCESS symbol that shows we must check if the road is clear - this creates a loop.

Loops are used in flowcharts to show when steps must be repeated.

In this example the loop shows we must keep checking for vehicles until the road is clear. Once the road is clear the answer to the question becomes YES and we can continue to the next PROCESS symbol and cross the road.

OUTPUT

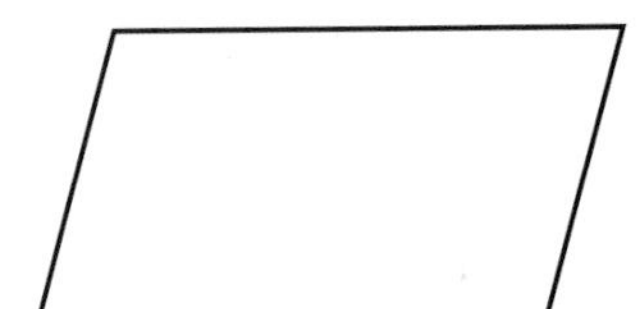

This symbol is used to show that a device must be turned ON or OFF.

OUTPUT symbols may have only one line going into and one line going out of them.

The flowchart below describes the problem of how to keep a room cool with a fan. OUTPUT symbols are used to turn the fan ON or OFF depending on the answer to the question in the DECISION symbol.

Tip:

This flowchart has no STOP symbol – the steps are repeated indefinitely.

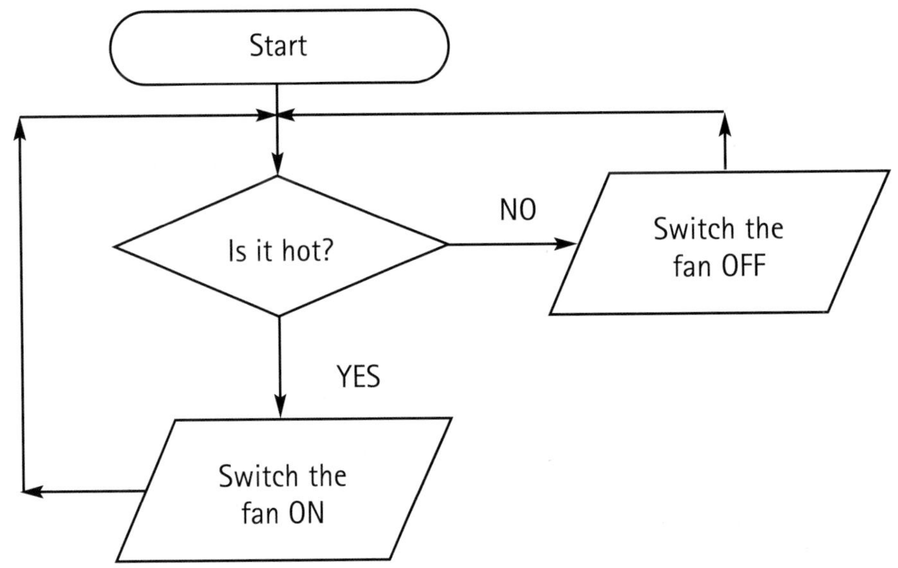

This flowchart uses a DECISION symbol to check if it is hot by asking a question. If the answer to the question is YES an arrow leads to an OUTPUT symbol that shows the fan must be switched ON to cool the room down. Another arrow leads from this OUTPUT symbol back to a place in the flowchart just before the DECISION symbol. This forms a loop to show the system must keep checking to see if it is hot. This loop will make the fan stay on until the answer to the question in the DECISION symbol is NO.

When the answer to the question is NO an arrow leads to a second OUTPUT symbol that shows the fan must be switched OFF. Another arrow leads from this OUTPUT symbol back to a place in the flowchart just before the DECISION symbol. This forms another loop to show that the system must keep checking to see if it is hot.

Rules for drawing flowcharts

When you draw flowcharts there are some rules, or conventions, that you should try to follow. These are:

- symbols should be arranged in the same order that the steps must be carried out;
- arrows connecting symbols should go down from top to bottom or across from left to right;
- arrows connecting symbols should not cross over.

In the flowchart below, none of these rules have been followed – can you work out what problem the flowchart is supposed to describe?

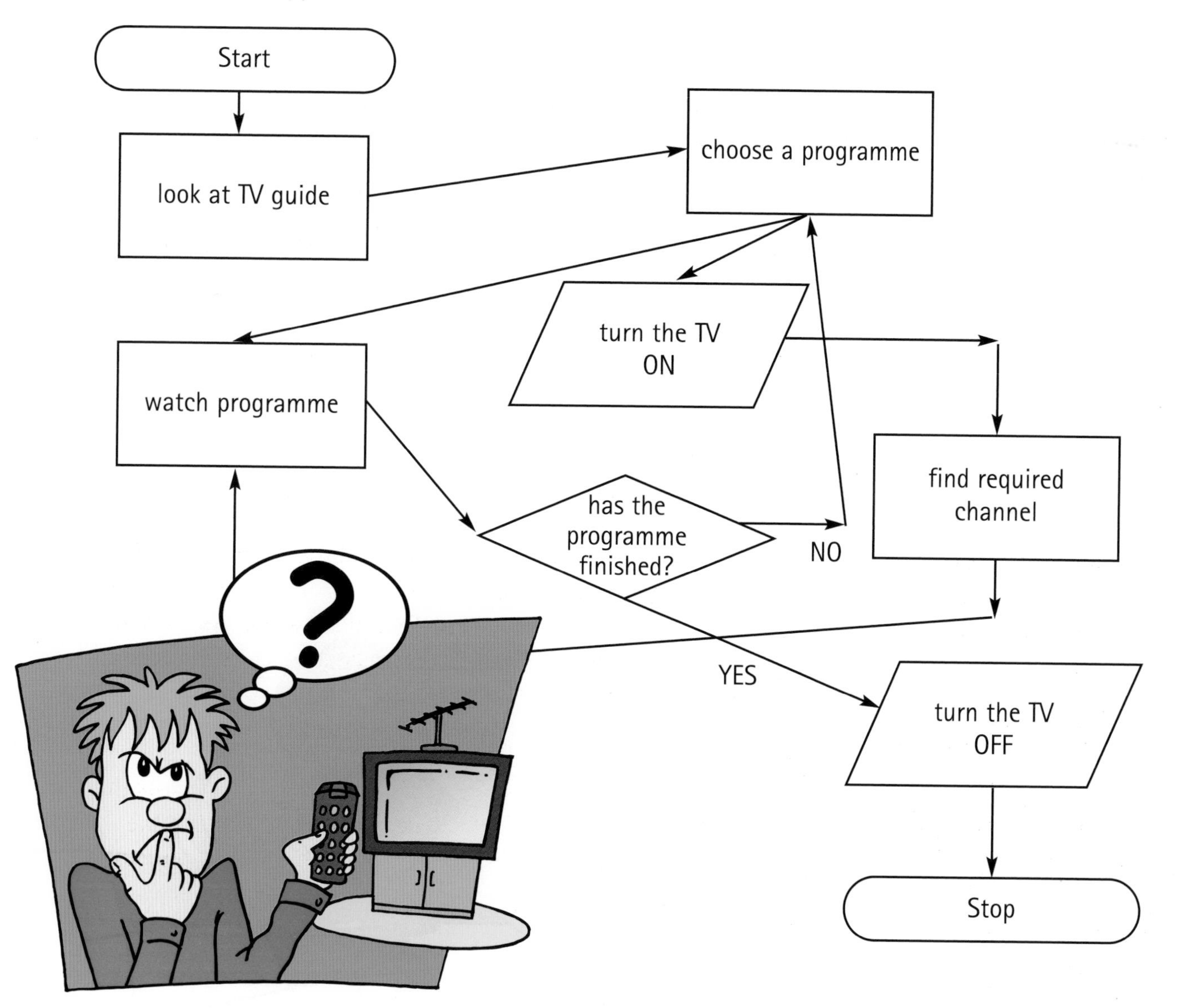

Did you work out that this flowchart describes the problem of choosing and watching a programme on TV?

Flowchart Exercises

(a) The steps you might work through when you get up in the morning before going to school are shown below.

Get out of bed

Pack school bag

Has the alarm gone off?

Get washed

Leave the house

Get dressed

Go back to sleep

Decide on the correct order for these steps.
Write the steps inside the blank flowchart symbols shown on Worksheet 1.

(b) Draw a flowchart to show the steps you might work through if you were feeding a dog. Use the space provided on Worksheet 1. To get you started, some of the steps you might include are listed below.

Is the dog hungry?

Open a tin of dog food

Put the food in the dog's bowl

Compare your finished flowchart with someone else's – is theirs different? Don't worry if it is – there's always more than one way to solve any problem.

(c) Draw a new version of the flowchart shown on the previous page. Use the space provided on Worksheet 1. Your new flowchart must follow the rules for drawing flowcharts.

Chapter 2 Sequences and Loops

In the last chapter you learned how to use flowcharts to describe the steps in a problem. In this chapter you will learn how to build flowcharts to control **simulated objects**. A simulated object is something that looks and works just like something in the real world but exists only inside the computer. The software we will use to do this is called **Flowol**. Flowol uses simulated objects called **mimics**. You are going to build a flowchart to control a set of traffic lights using a Flowol mimic.

Getting started - planning

Make sure you have a copy of Worksheet 2.1.

The first thing you need to do is work out the sequence of events at a set of traffic lights when the lights change to make vehicles stop or go. To help with this you are going to watch a PowerPoint presentation called trafficlights. Your teacher will tell you where to find this file. If you are working on your own, the presentation can be downloaded from www.payne-gallway.co.uk/flowol.

Run the trafficlights presentation.

Watch the traffic lights changing.

Write the light sequence on Worksheet 2.1 in the first table labelled Traffic light sequence – the first two steps have been done for you.

Traffic light sequence

Red light **ON**
Red light and amber light **ON**

 When you have finished press ESC to stop the presentation.

Ask your teacher to check the steps you've written down before carrying on.

So far, we've thought about the lights sequence but that's not all that we need to do. When you watched the traffic lights did they change straight away, or were there times when there was pause or delay? Think about when the red light came on – did it stay on for a certain amount of time before the amber light was turned on?

All traffic lights need delays built in to allow enough time for vehicles to move through them. Some traffic lights take the number of vehicles on the road into account and use different delays at quiet or busy times of day.

 Run the trafficlights presentation again - this time concentrate on the delays built into the light sequence.

 Count the number of seconds any delay lasts for – you don't have to be exact – we just need to know where the delays are and approximately how long they last.

 Write the light sequence down again on Worksheet 2.1 in the second table labelled Traffic light sequence with delays. Include the delays this time. To help you get started the first few steps have been completed for you.

Traffic light sequence with delays

Red light **ON**
DELAY for **30 seconds**
Red light **ON** and amber light **ON**

 When you have finished press ESC to stop the presentation.

Ask your teacher to check the steps you've written down before carrying on.

Now that you have thought about the sequence of events at the traffic lights you are ready to load Flowol and build a flowchart to control them. You can load the Flowol program in one of two ways:

 *Either* double-click the Flowol icon on your desktop,

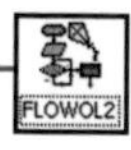

 *or* click Start at the bottom left of the screen, then click Programs, then click

Next we need to load a mimic called bridligh – follow the steps below to do this.

 Click Window, Mimic.

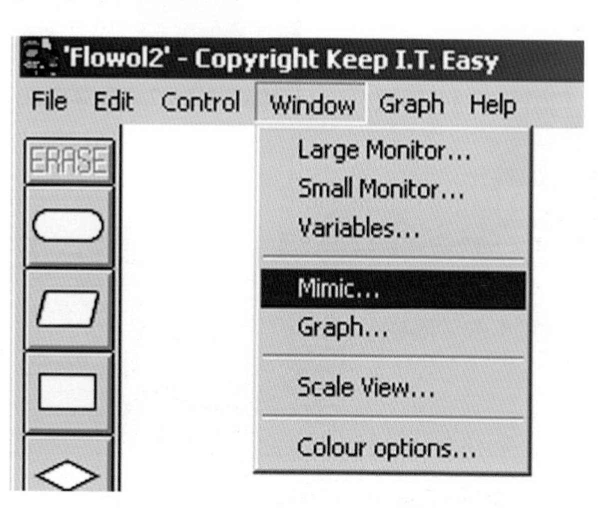

Click on bridligh in the list of mimics.

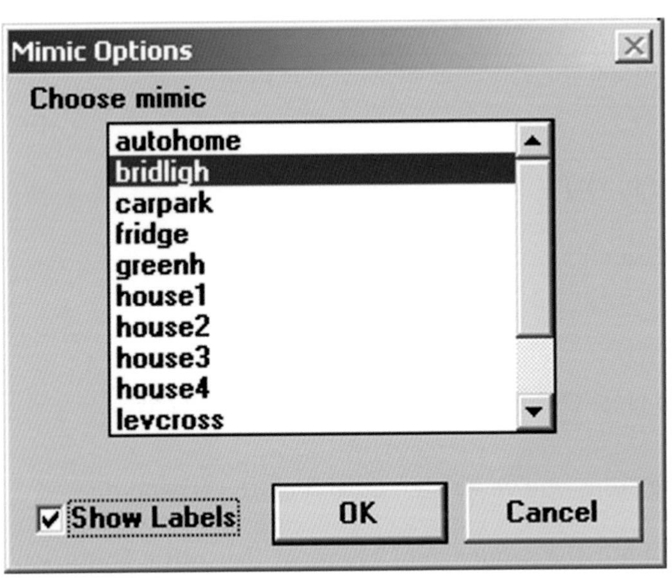

Click the Show Labels box.

This will label the inputs and outputs on the mimic.

Click OK.

Flowol will load the bridligh mimic.

Next we'll spend a few minutes finding out about the outputs on this mimic.

The three outputs labelled out 1, out 2 and out 3 are connected to the traffic lights on this side of the bridge.

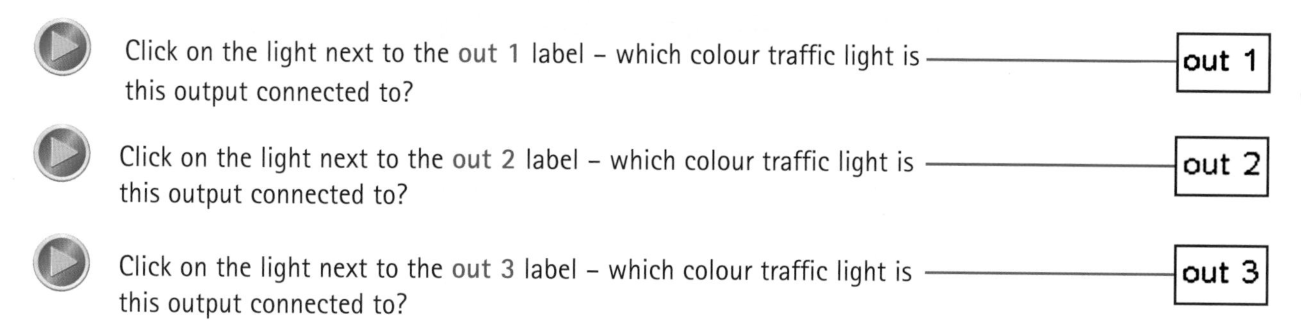

Click on the light next to the out 1 label – which colour traffic light is this output connected to?

Click on the light next to the out 2 label – which colour traffic light is this output connected to?

Click on the light next to the out 3 label – which colour traffic light is this output connected to?

Now you know more about outputs on the mimic we'll build a flowchart to control the traffic light sequence on this side of the bridge.

Building a flowchart

The first symbol we need in any flowchart is START.

Click on the START/STOP symbol on the left of the screen.

Move the mouse pointer across to the right of the screen – you will see a START/STOP symbol being dragged along with the mouse.

Click the left mouse button.

Flowol will put a START/STOP symbol on the screen like the one below.

Next we need to tell Flowol that we want this to be a START symbol.

Click the Start button at the bottom of the screen. **Start**

The word Start will appear inside the START/STOP symbol.

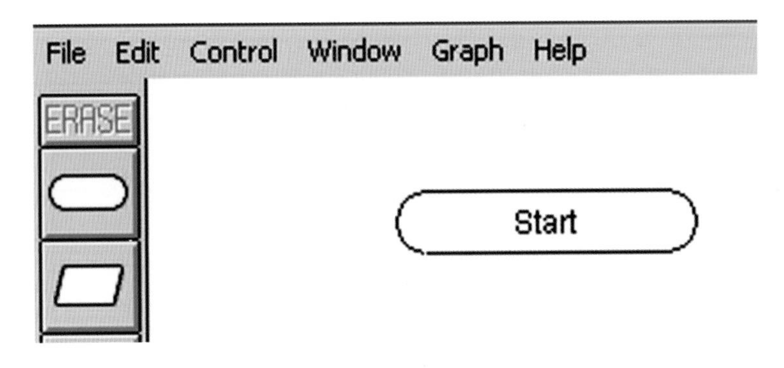

That's the first symbol in our traffic lights flowchart taken care of!

The next symbol we need to add is an OUTPUT symbol. This will be used to turn the red light ON. Follow the steps below to add this symbol.

- Click on the OUTPUT symbol on the left of the screen.
- Move the mouse pointer across to the right of the screen – you will see an OUTPUT symbol being dragged along with the mouse.
- Click the left mouse button when you are just underneath the START symbol. Flowol will put an OUTPUT symbol on the screen like the one below.

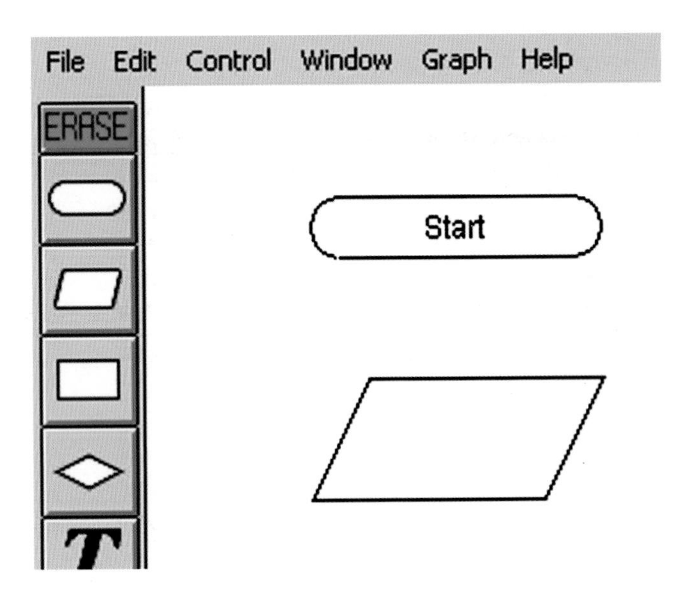

Don't worry if you see the message shown below – this means you've put the OUTPUT symbol too close to the START symbol. If this happens just click OK and try again.

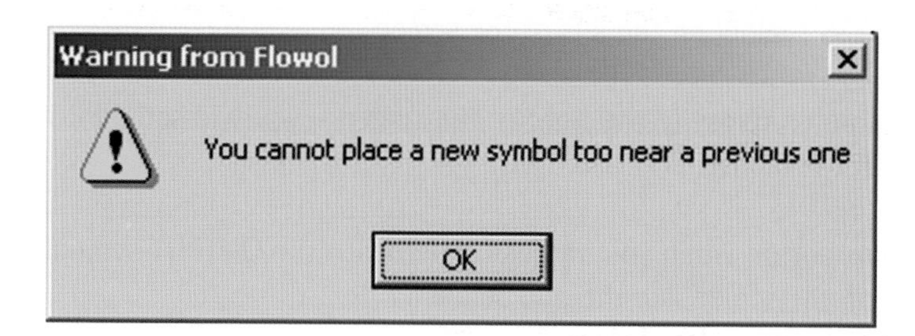

2

Now we need to tell Flowol what we want this output symbol to do. Look at the first table on Worksheet 2.1. You will see the first light in the sequence is red. The red light on this side of the bridge is labelled out 3 so this output symbol must turn output 3 ON. Follow the steps below to do this.

out 3

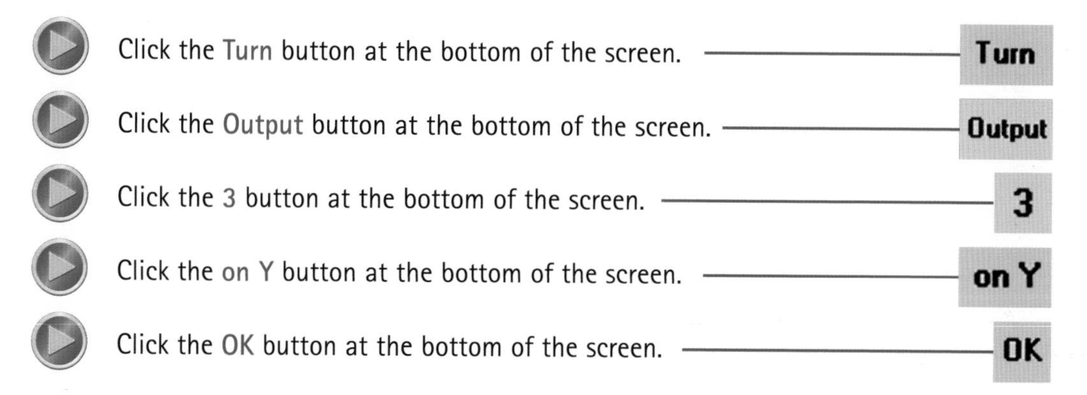

- Click the Turn button at the bottom of the screen. — Turn
- Click the Output button at the bottom of the screen. — Output
- Click the 3 button at the bottom of the screen. — 3
- Click the on Y button at the bottom of the screen. — on Y
- Click the OK button at the bottom of the screen. — OK

Your flowchart should now look like this:

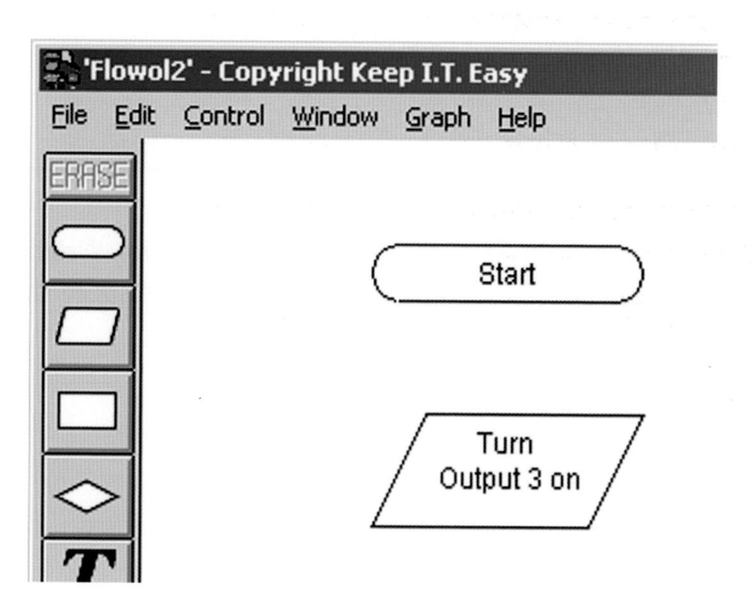

Look at the second table on Worksheet 2.1. You will see that after the red light is turned on a delay is needed. To set a delay we must use a PROCESS symbol.

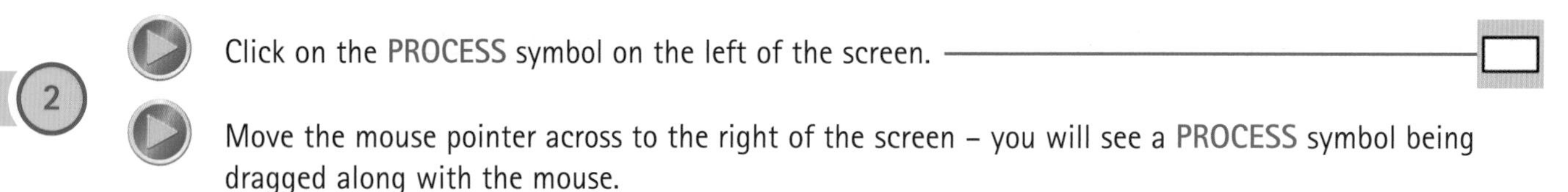

- Click on the PROCESS symbol on the left of the screen.

- Move the mouse pointer across to the right of the screen – you will see a PROCESS symbol being dragged along with the mouse.

- Click the left mouse button when you are just underneath the OUTPUT symbol. Flowol will put a PROCESS symbol on the screen like the one below.

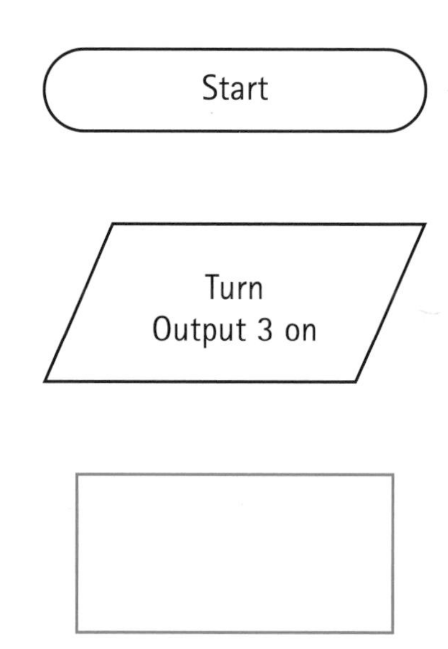

Now we need to use the Edit Process box at the bottom of the screen to tell Flowol that this process symbol is going to be a delay.

- Click the Delay button at the bottom of the screen.

Next we need to tell Flowol how many seconds we want the delay to last for. We're not going to set this delay to the value you've written down – we'll use a 5-second delay to speed things up when the flowchart is running.

Click the 5 button at the bottom of the screen to set a five second delay. — **5**

The Edit Process box at the bottom of your screen should look like the one below.

2

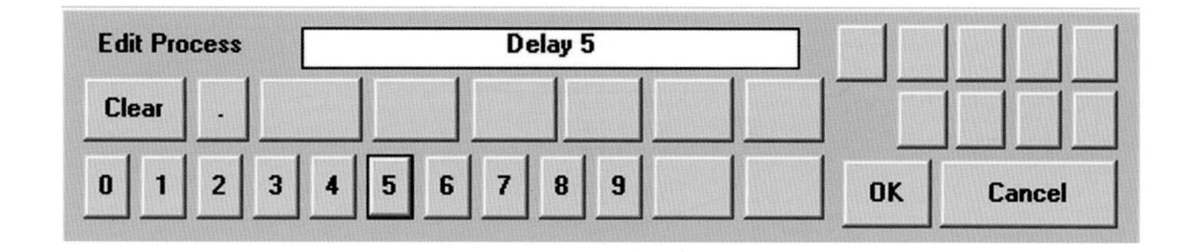

If you've made a mistake just click on the Clear button and repeat the steps above. — **Clear**

Click the OK button when you've finished. — **OK**

Your flowchart should now look like the one below.

Start

Turn
Output 3 on

Delay 5

Tip:

If you make a mistake and want to correct or erase one of your flowchart symbols, right-click it. It will turn red and you can then erase or edit it.

Look at the second table on Worksheet 2.1. You will see that after the first delay the next lights in the sequence are red and amber together. The red light is already on so the next step in the flowchart just needs to turn the amber light ON. The amber light on this side of the bridge is labelled out 2 so this output symbol must turn output 2 ON. Follow the steps below to do this. — out 2

Put another OUTPUT symbol underneath the Delay 5 process symbol.

 Set this output symbol to turn output 2 ON using the Edit Output box at the bottom of the screen.

Your flowchart should now look like the one below.

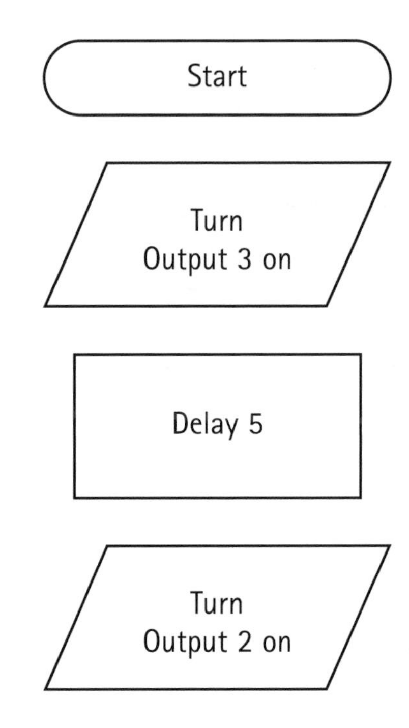

Look at the second table on Worksheet 2.1. You will see that after the amber light is turned on another delay is needed. To set this delay we must use another PROCESS symbol.

 Put another PROCESS symbol underneath the Turn Output 2 on output symbol. 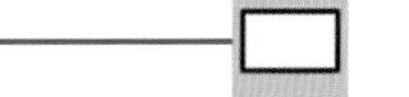

 Set this process symbol to give the length of delay you need (e.g. 3) using the Edit Process box at the bottom of the screen.

The next steps in the sequence must turn the red and amber lights OFF before turning the green light ON.

The red and amber lights are labelled out 3 and out 2 so we could use two output symbols to turn output 3 and output 2 OFF.

The green light is labelled out 1 so we could use a third output symbol to turn output 1 ON.

out 3

out 2

out 1

Flowol will let us use a single output symbol to turn more than one output ON or OFF at the same time. We can complete these two steps using just one output symbol, making our flowchart more efficient.

Being able to build flowcharts that are efficient because they use as few symbols as possible will show that your computer control skills have improved. Follow the instructions below to do this.

2

Put another OUTPUT symbol underneath the Delay 3 PROCESS symbol.

In the Edit Output box at the bottom of the screen:

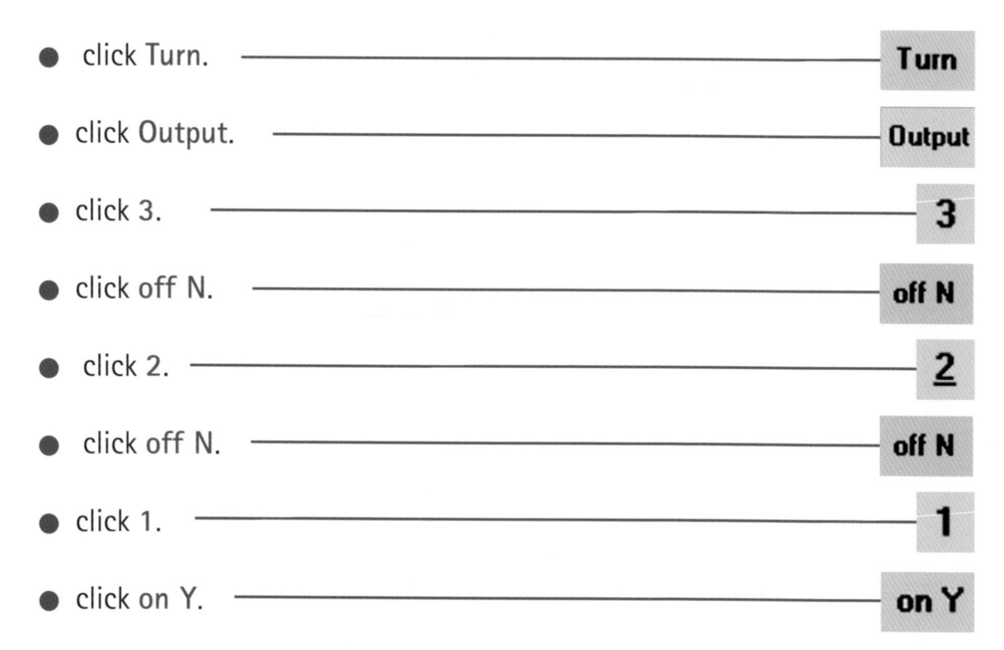

- click Turn.
- click Output.
- click 3.
- click off N.
- click 2.
- click off N.
- click 1.
- click on Y.

The Edit Process box at the bottom of your screen should look like the one below.

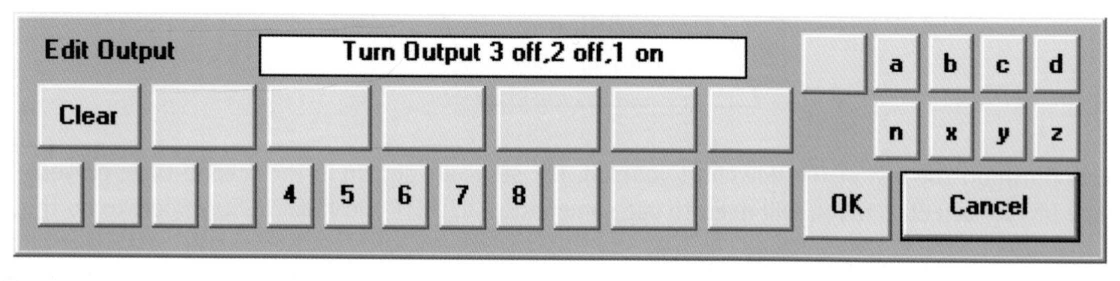

If you've made a mistake just click the Clear button and repeat the steps above.

Click the OK button when you've finished.

Your flowchart should now look like the one below.

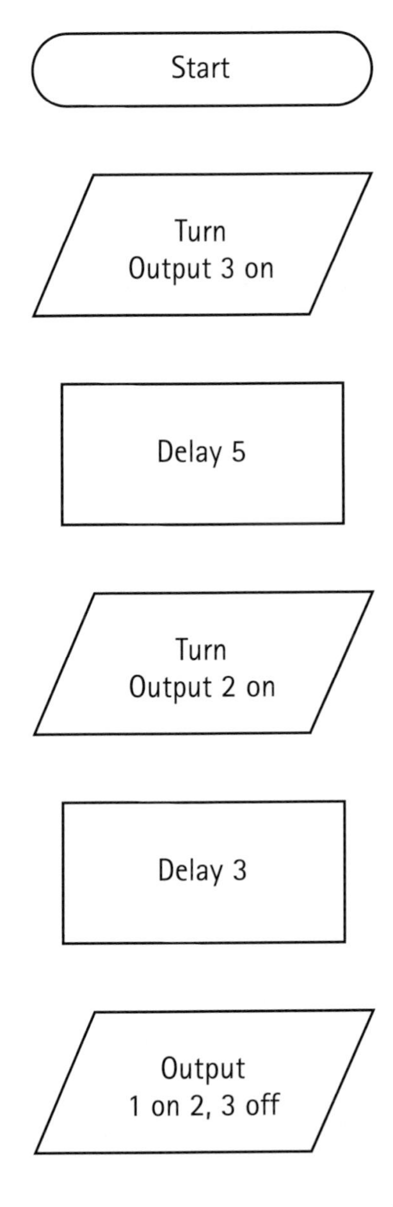

Now finish off the rest of the flowchart yourself. The steps are shown in the second table on your copy of Worksheet 2.1. You will need to use some more OUTPUT and PROCESS symbols to do this.

If you're not sure what to do now or you haven't completed Worksheet 2.1 ask your teacher for a copy of the bridge lights help sheet.

Joining the symbols

The last thing we need to do before testing this flowchart is to join the symbols together. The arrow buttons on the left of the screen are used to do this.

We'll look at what the YES and NO arrow buttons are used for in the next chapter. For now the button we need to use is the ordinary Arrow button (referred to as the Always Line in the prompt at the bottom of the Flowol screen when you click it).

We'll start by joining the first two symbols on the flowchart together.

Click the Arrow button.

Click on the START symbol at the beginning of your flowchart.

Click on the OUTPUT symbol underneath the START symbol.

A line will appear joining the first two symbols on the flowchart together.

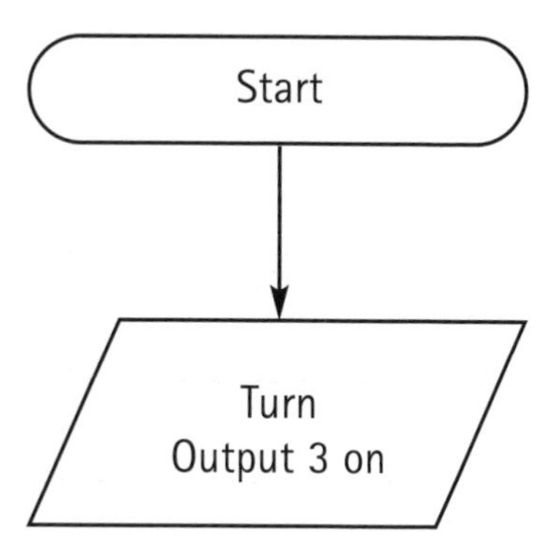

Now carry on and join up the other symbols in your flowchart.

Testing the flowchart

Now the symbols are joined together we can try running the flowchart.

Click the Run button on the left of the screen. — Run

Flowol will run the flowchart. You will see the traffic lights changing but the error message below will appear when the last symbol in the flowchart has been reached.

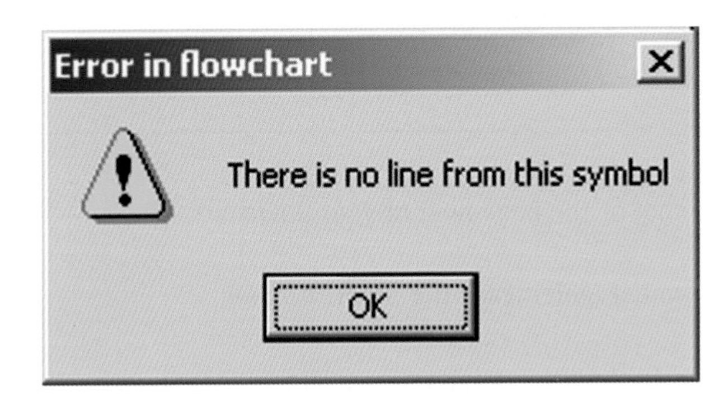

This error message has appeared because flowcharts must have a STOP symbol at the end or an arrow leading back from the last symbol to another place in the flowchart.

If we put a STOP symbol at the end of this flowchart the lights will run through the sequence once then stop – this isn't what we want to happen! The traffic lights must keep running through the sequence forever to control the flow of vehicles over the bridge. Have you ever seen a set of traffic lights change once then turn off? To make the traffic lights repeat the same sequence over and over again we need to use a loop.

In the last chapter you learnt that a loop in a flowchart is shown by an arrow leading back to another place in the flowchart. To make this traffic light sequence repeat we need to create a loop leading from the last symbol in the flowchart back to the top. This will make the light sequence repeat over and over again without stopping. Follow the steps below to do this.

Click the Arrow button. — →

Click on the last symbol at the bottom of your flowchart.

Use the scroll bar on the right of the screen to move back to the top of your flowchart.

Click the Turn Output 3 on output symbol.

A line will appear joining the symbol at the bottom of your flowchart to the top of the flowchart. This has created a loop that will repeat the light sequence by going back to the beginning of the flowchart after the step in the last symbol has been carried out.

Now we can try running the flowchart again.

Click the Run button on the left of the screen.

Run

Flowol will run the flowchart. You will see the traffic lights changing but this time the light sequence will repeat over and over again.

Stop the flowchart from running by clicking the Stop button on the left of the screen.

Correcting the flowchart

If your flowchart doesn't work follow the steps below to check and correct it.

Use your copy of Worksheet 2.1 to check your flowchart and make sure that:

- all the symbols are in the correct order;
- the right outputs are being turned ON or OFF inside each DECISION symbol;
- the delays last for the right amount of time.

If you need to change what's inside a symbol:

- Click on the Hand tool.
- Click on the symbol you want to change.
- Click on Clear in the Edit box at the bottom of the screen.
- Use the Edit box to correct what's inside the symbol then click OK.

Clear

OK

If you need to delete a line or a symbol and draw it again:

- Click on the Hand tool.
- Click on the line or symbol.
- Click on the Erase button.
- Draw the line or symbol again.

ERASE

If you need to move a symbol:

- Delete all the lines connected to the symbol – follow the steps above to do this.
- Click on the Hand tool, then click on the symbol.
- Click on and hold down the left mouse button.
- Drag the symbol to its new position then let go of the left mouse button.

Tip:

If you have missed out a flowchart symbol, you can move the symbols below it down to make room for the missing one.

When you've finished run your flowchart again – everything should work this time! If your flowchart still doesn't work ask your teacher to help you check through it.

If you want to save your flowchart before going any further, look ahead a few pages to the end of this chapter for how to do this. Otherwise, continue on now with Task A, which develops this flowchart further.

Task A

Before you start this task you need a copy of Worksheet 2.2.

So far we've only dealt with half of this problem. There are two sets of traffic lights – one on each side of the bridge – controlling the vehicles on both sides. The lights must work together so that vehicles don't cross over the bridge at the same time and crash. To solve the other half of this problem we must create another flowchart to control the traffic lights on the other side of the bridge.

To get started look at your copy of Worksheet 2.2 – you will see a table like the one shown below.

Traffic lights - set 1	Traffic lights - set 2
Red light **ON**	Red light **ON**
Red light **ON** and amber light **ON**	Red light **ON**
Green light **ON**	Red light **ON**
Amber light **ON**	Red light **ON**
Red light **ON**	Red light **ON**
Red light **ON**	

The light sequence for the traffic lights on this side of the bridge is shown in the first column of the table labelled Traffic lights – set 1. You've already worked out this sequence and created a flowchart for it. To prevent cars from crashing, the sequence for the traffic lights on the other side of the bridge must be the reverse of the sequence shown in this column.

Write the sequence for the traffic lights on the other side of the bridge in the second column of the table labelled Traffic lights – set 2.

Ask your teacher to check the steps you've written down before carrying on.

Now you're ready to build a flowchart to control this light sequence. Put this flowchart next to the first flowchart. When you've finished and are ready to test your solution both flowcharts will run together.

 To get started, put a START symbol next to the START symbol in the first flowchart.

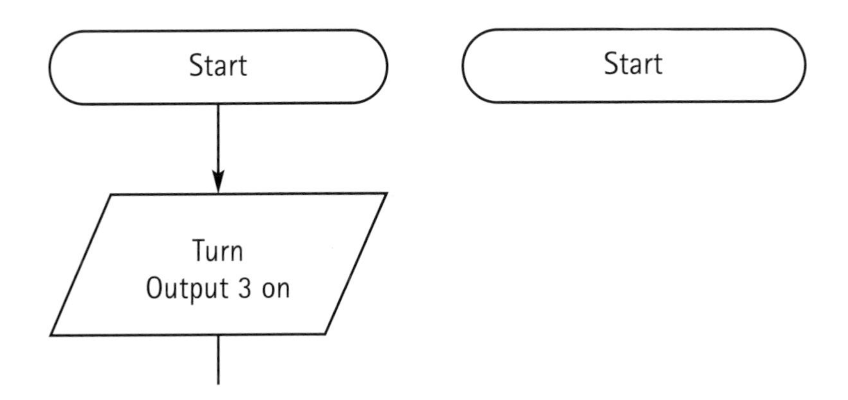

When you've finished

The first thing you'll need to do is print your flowchart. Before doing this it's a good idea to label the flowchart with your name. Follow the steps below to do this.

 Click on the Label button on the left of the screen. 

 Click at the top of the screen above your flowchart.

 Type your name – it will appear in the Edit label box at the bottom of the screen.

 Click Medium and OK.

Your name will appear at the top of the flowchart like the example shown below.

John Wheat

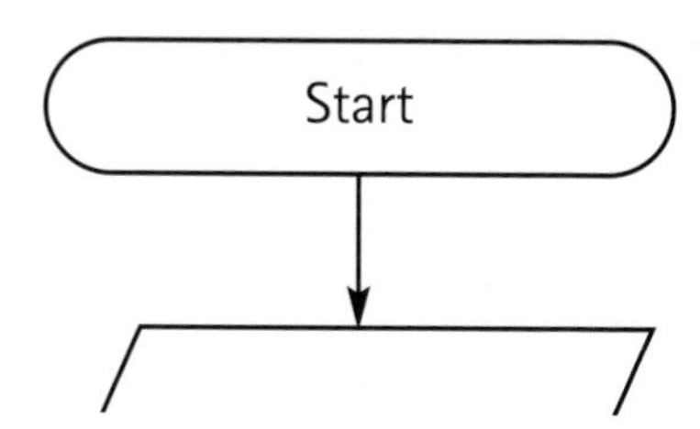

Click File, Print on the main menu at the top of the screen.

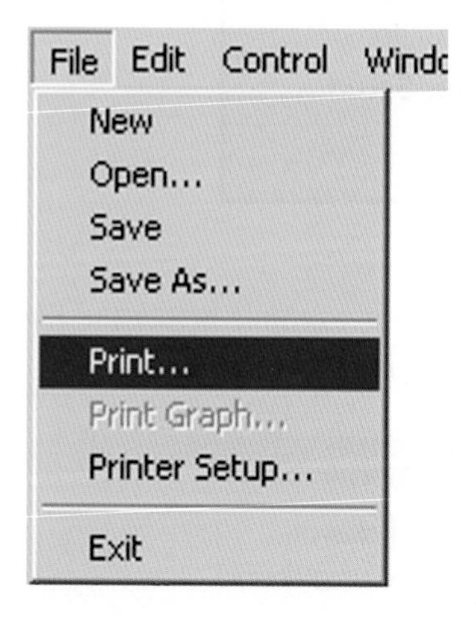

A window like the one below will appear.

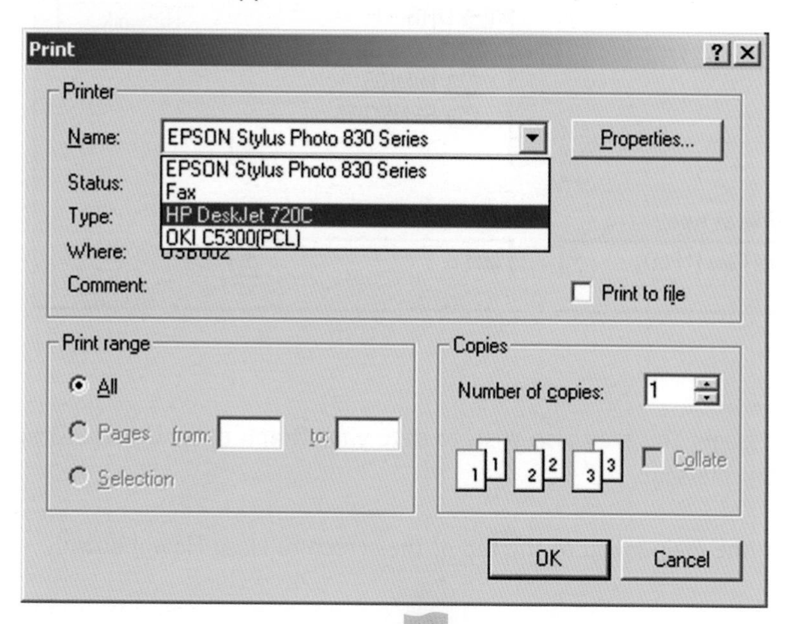

Choose a printer and then click OK.

Next follow the steps below to save the flowchart and close Flowol down.

Click File, Save on the main menu at the top of the screen.

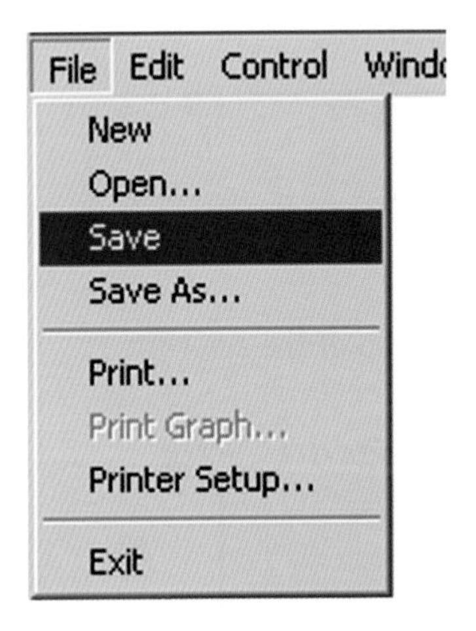

Type a file name for your flowchart. In the example below the file name bridge has been entered. (Flowol will allow only up to 8 characters in a file name.)

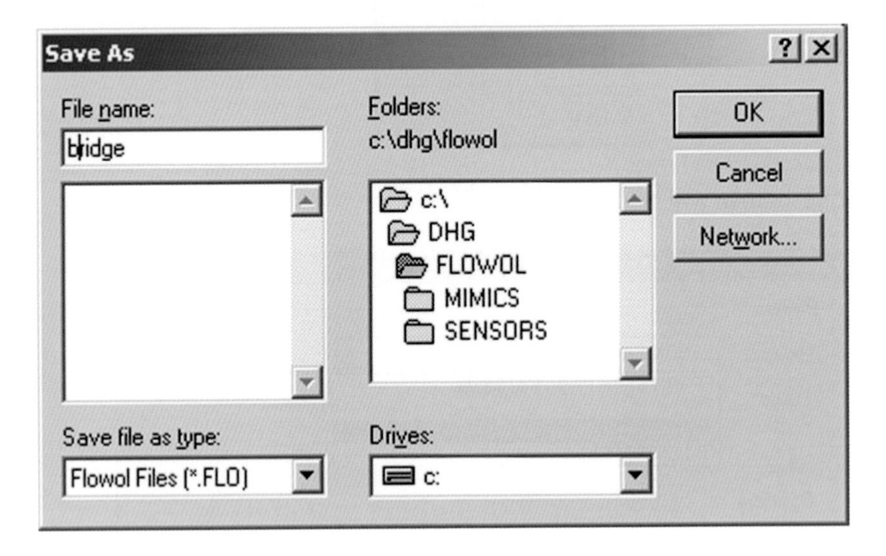

Click OK.

You will see a message: You may also wish to save your Graph data. Use the Graph menu. Click OK. You can ignore this message.

Click File, Exit on the main menu at the top of the screen to close Flowol down.

Chapter 3 Making Decisions

In the last chapter you learned how to repeat the same steps in a flowchart using a simple loop. To control most devices, computers need to know what is happening in the world around them. Computers detect this information using input signals from **sensors**. Sensors can send two types of input signal, called **digital** and **analogue**, to a computer. In this chapter you will learn how input from sensors can be used to control loops by asking questions.

Before we get started make sure you have a copy of Worksheet 3.

3

Getting started – planning

Load Flowol.

You need to load a mimic called fridge – follow the steps below to do this.

Click Window, Mimic.

Click on fridge in the list of mimics.

 Click the Show Labels box and click OK.

Flowol will load the fridge mimic – your screen will look like the one below.

Before we start building a flowchart to solve this problem we'll spend a few minutes finding out about the inputs and outputs on the mimic.

This mimic has one input from a sensor labelled in 1. This is a switch that sends out a digital signal. A digital signal can only have two values - either ON or OFF. When the fridge door is closed the switch is OFF. When the fridge door is open the switch is ON. When our flowchart is running we'll simulate the opening and closing of the fridge door by clicking on the white switch underneath the in 1 label.

in 1

in 1

The mimic has one output labelled out 1 connected to the light inside the fridge. Sending a signal to this output will turn the fridge light ON or OFF.

out 1

Now you know a little more about this mimic we can think about the control problem that needs solving.

The fridge light must be OFF when the fridge door is closed.

The fridge light must come ON when the fridge door is opened.

The light must stay ON until the fridge door is closed again.

This problem needs to use a loop to check continuously to see if the fridge door is open or closed. To control this loop we must ask the question "Is the fridge door open?" If the answer to this question is YES the fridge light must be turned ON. If the answer to this question is NO the fridge light must be turned OFF.

A flowchart to describe this problem is shown below.

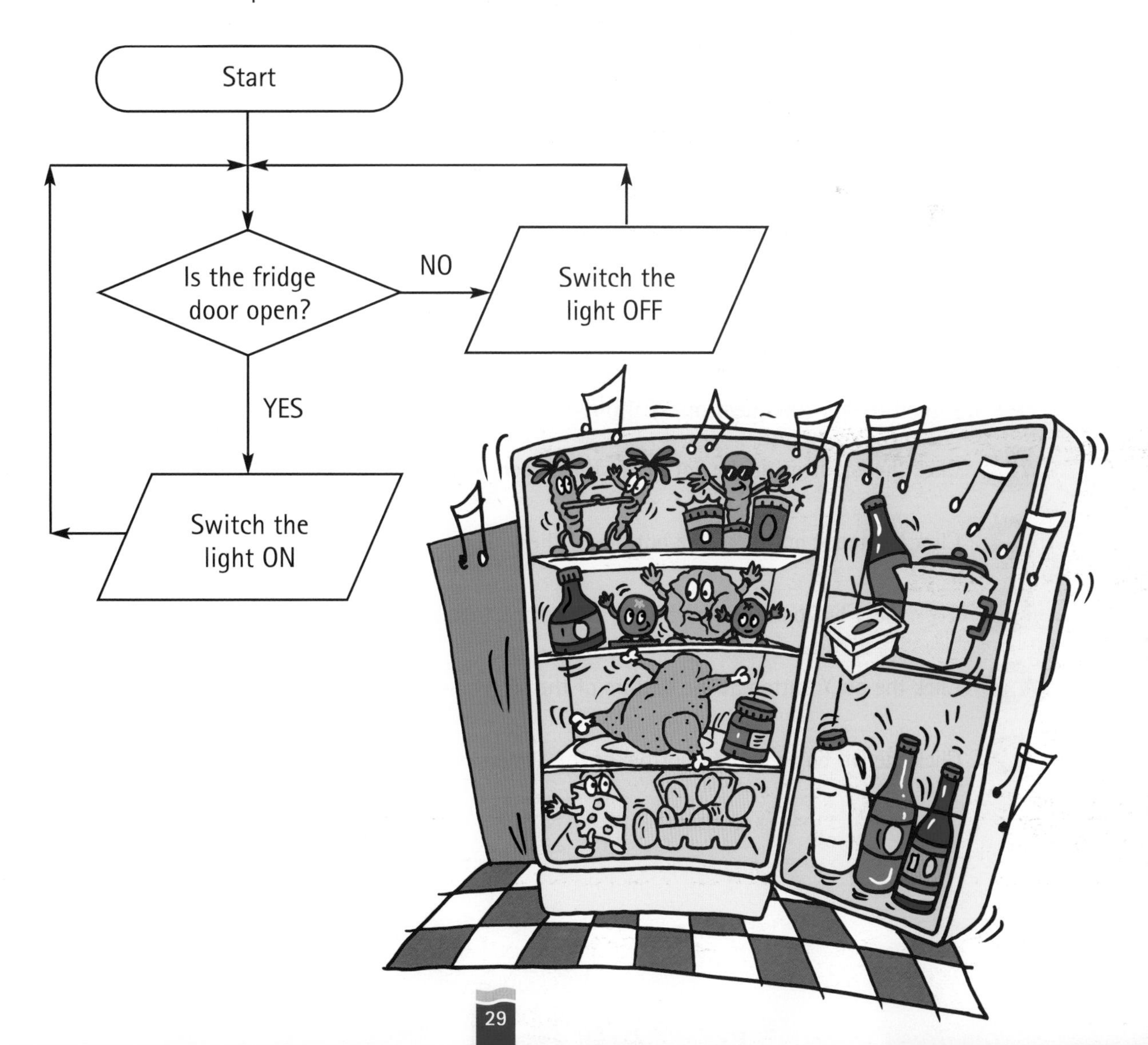

Building the flowchart

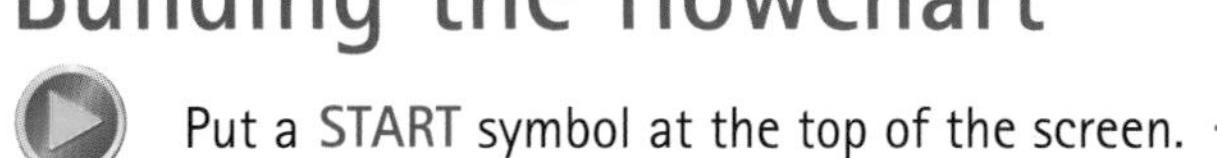

Put a START symbol at the top of the screen.

The next symbol we need to add is a DECISION symbol. We are going to use this symbol to see if the fridge door is open or closed by asking a question that checks the value of the sensor inside the door labelled in 1.

in 1

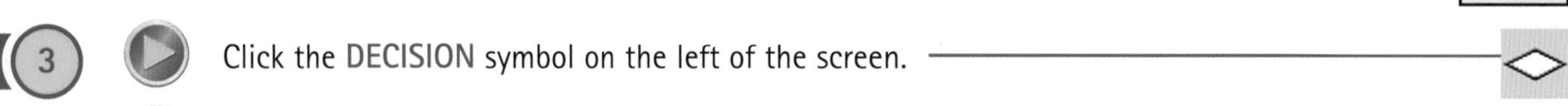

3

Click the DECISION symbol on the left of the screen.

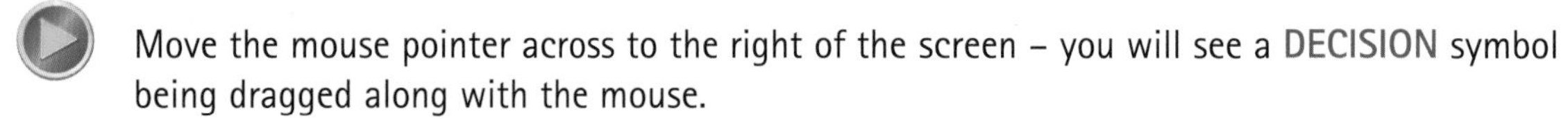

Move the mouse pointer across to the right of the screen – you will see a DECISION symbol being dragged along with the mouse.

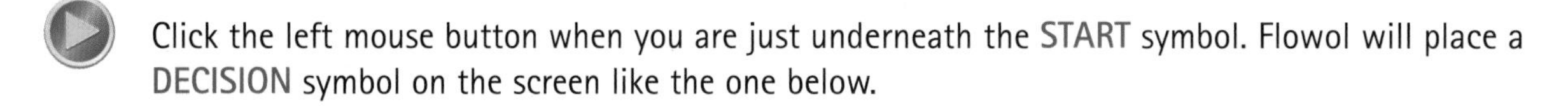

Click the left mouse button when you are just underneath the START symbol. Flowol will place a DECISION symbol on the screen like the one below.

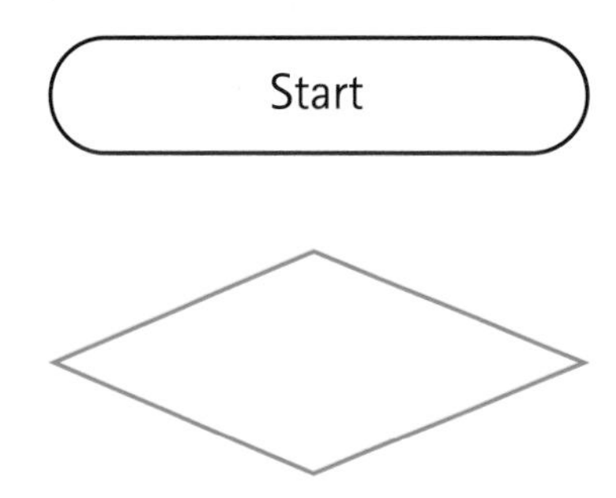

In Flowol, decision symbols are used to control loops by asking questions that check inputs. In this case we need to ask the question, Is input 1 on?

Click the Is button at the bottom of the screen. — **Is**

Click the Input button at the bottom of the screen. — **Input**

Click the 1 button at the bottom of the screen. — **1**

Click the on Y button at the bottom of the screen. — **on Y**

Click the OK button at the bottom of the screen. — **OK**

Your flowchart should look like the one below.

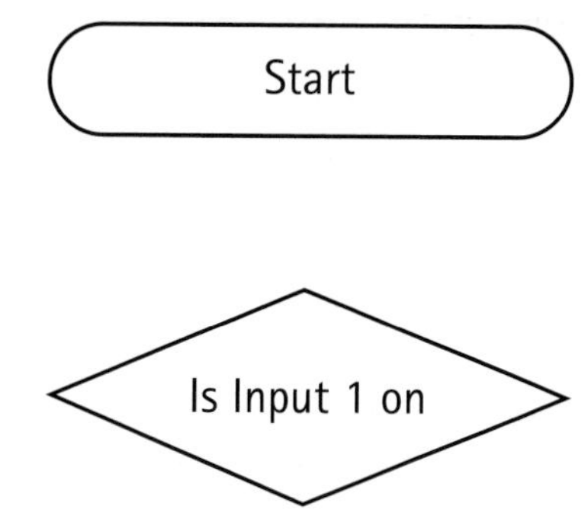

Now that a decision symbol has been added to the flowchart, we must say what should happen next. There are always just two possible answers to the question inside a decision symbol – YES or NO. For this problem if the answer to the question is YES the fridge light must be turned ON. To do this we need to use an OUTPUT symbol to turn output 1 ON.

Put an OUTPUT symbol on the right side of the decision symbol.

Set this output symbol to turn output 1 ON.

Your flowchart should look like the one below.

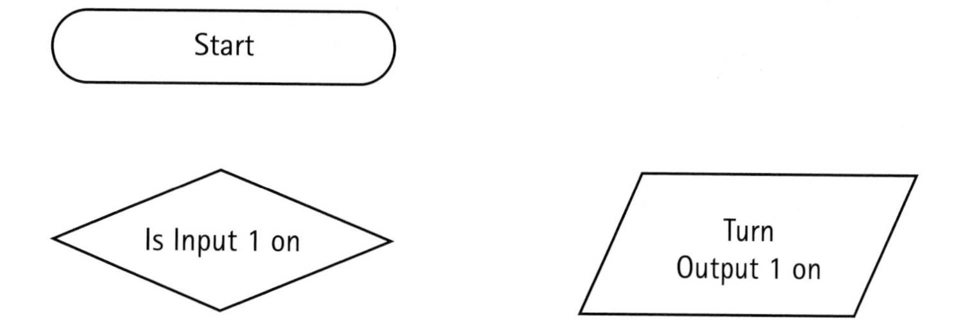

Next we'll join these first few symbols up.

Use the Arrow button on the left of the screen to join the START symbol at the beginning of the flowchart to the DECISION symbol underneath.

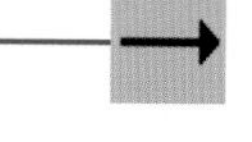

Click the Yes Arrow button on the left of the screen.

 Click the DECISION symbol.

 Click the Turn Output 1 on symbol.

A YES arrow will appear.

Your flowchart should look like the one below.

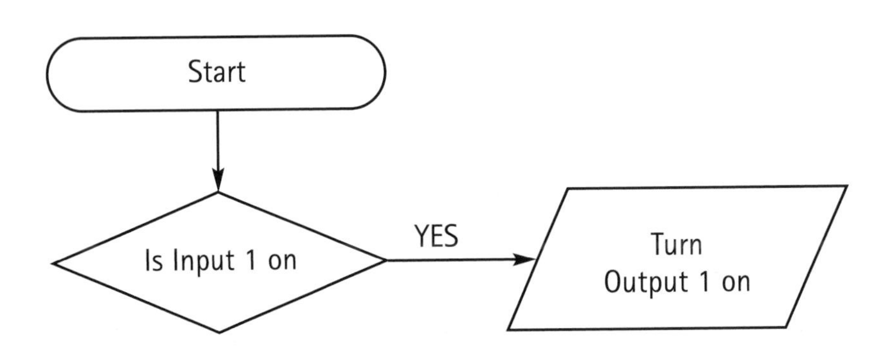

Next we need to think about what should happen when in 1 is OFF making the answer to the question inside the decision symbol NO. This tells us the fridge door is closed so we must turn the light inside OFF. To turn the light OFF we will need to use another OUTPUT symbol.

in 1

 Put another OUTPUT symbol underneath the decision symbol.

 Set this symbol to turn output 1 OFF.

Now we need to join this output symbol to the decision symbol above with a NO arrow.

NO

 Click the NO Arrow button on the left of the screen.

 Click the DECISION symbol in the flowchart.

 Click the Turn Output 1 off symbol underneath.

A NO arrow will appear.

Your flowchart should now look like the one below.

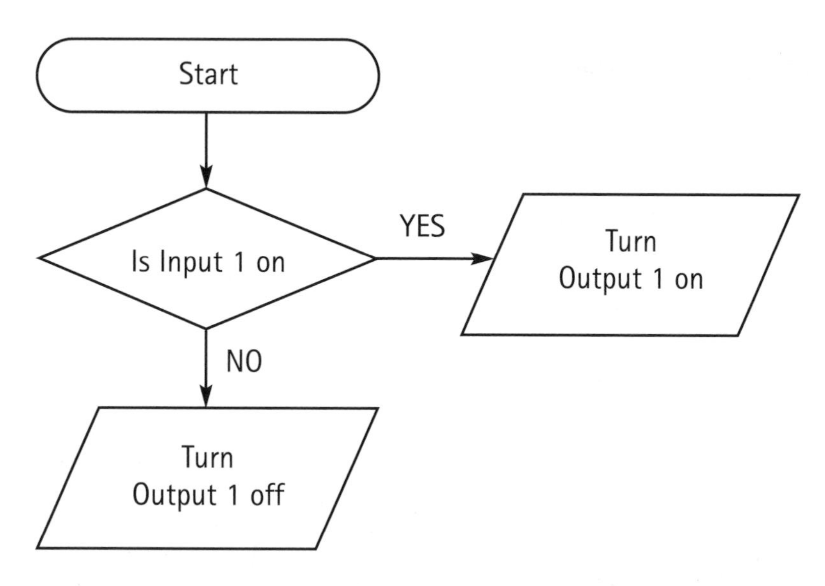

The last thing we need to do is add lines from both output symbols to join back to the top of the flowchart. This will complete the loops that will make the system check continuously to see if the fridge door is open.

First we'll join the **Turn Output 1 on** output symbol back up to the top of the flowchart.

- Click the **Arrow** button on the left of the screen.
- Click the **Turn Output 1 on** output symbol.
- Click just above this symbol.

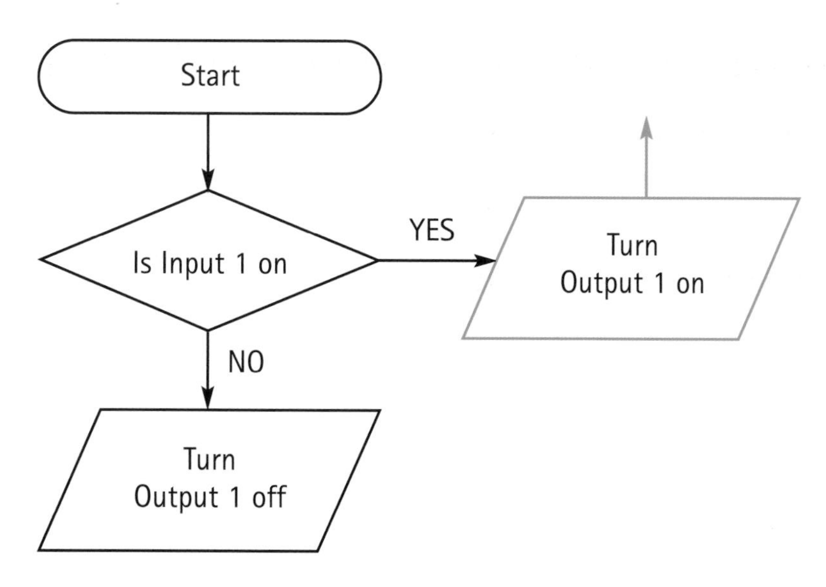

Click against the arrow joining the first twosymbols at the top of the flowchart together.

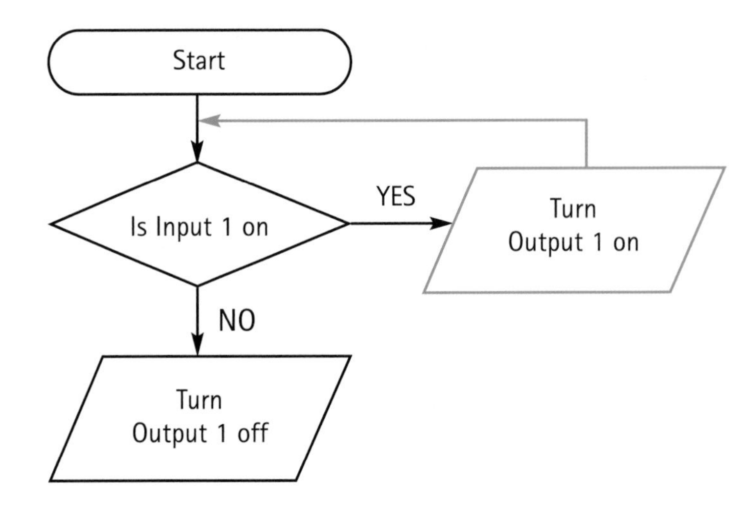

Click inside the DECISION symbol to make the lines join together.

Your flowchart should look like this.

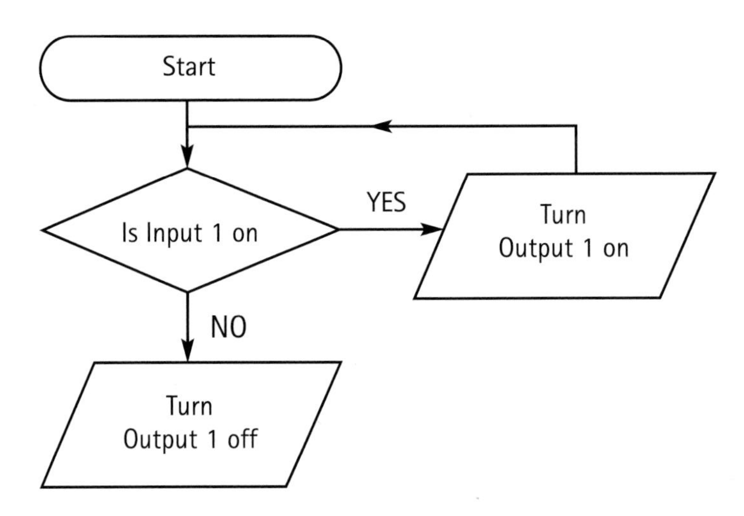

Now join the Turn Output 1 off output symbol back to the top of the flowchart. When you have done this your flowchart should look like the one on the right.

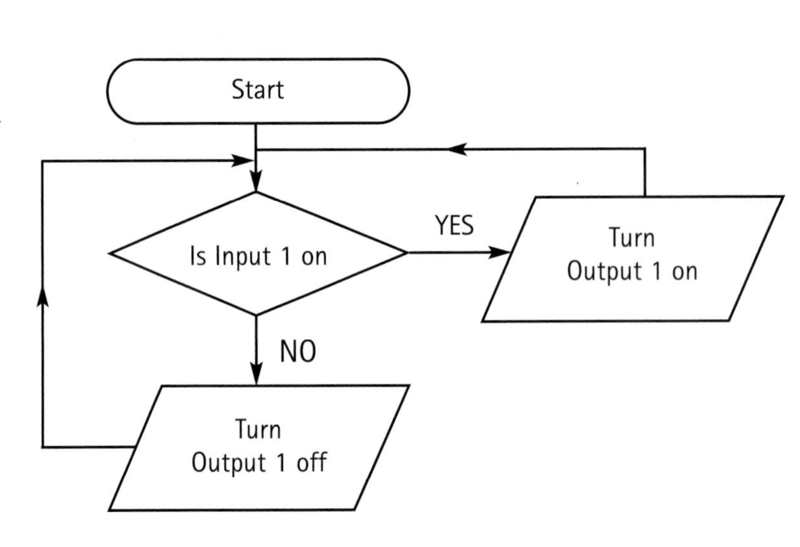

If the lines you draw don't join up in the right place use the Erase button to remove them and draw them again. ERASE

When all the symbols are joined together we can test the flowchart by running it and seeing what happens.

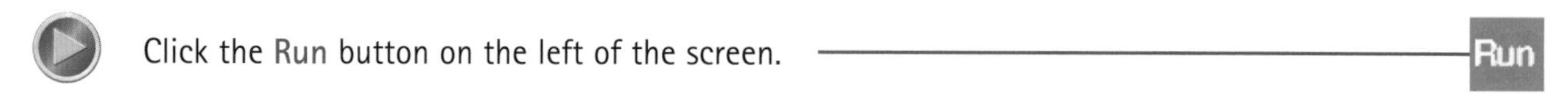

Click the Run button on the left of the screen. Run

Flowol will run the flowchart.

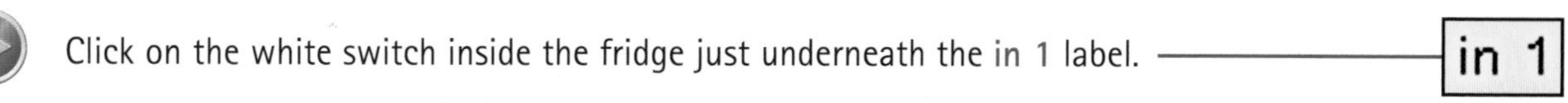

Click on the white switch inside the fridge just underneath the in 1 label. in 1

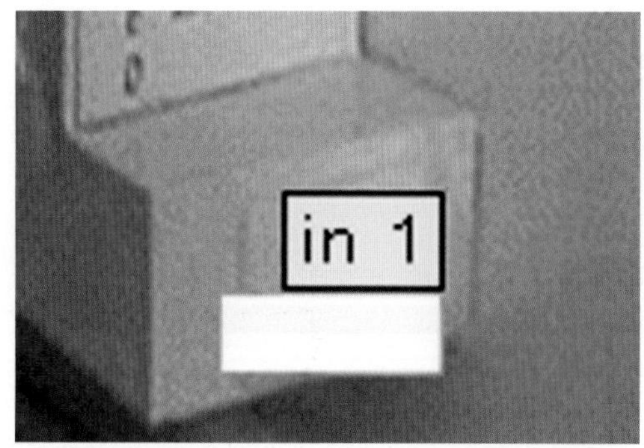

You should see the light in the fridge turning on and off. By clicking on this switch you are simulating the fridge door being opened and closed.

If your flowchart isn't working, check that it is the same as the one shown on the previous page and correct any mistakes before trying again.

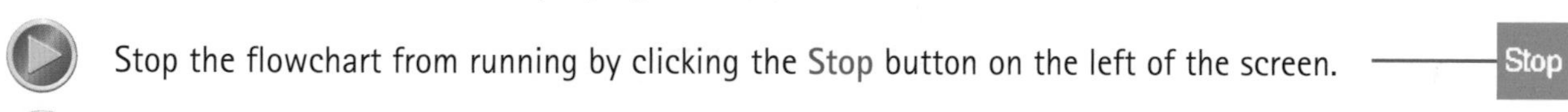

Stop the flowchart from running by clicking the Stop button on the left of the screen. Stop

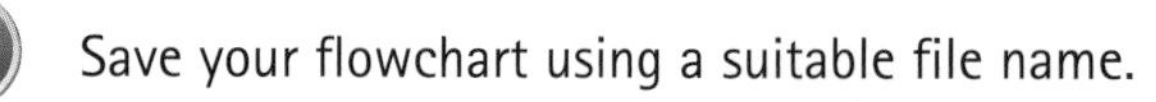

Save your flowchart using a suitable file name.

Task B

You are going to create a flowchart of your own to solve a similar problem to turning the fridge light on and off. This time the problem is about monitoring the light level outside a house and turning some security lights on when it is dark.

First you need to load a mimic called house1.

From the main menu select File, New.

You may see a message: Are you sure you want to erase your data? Click OK.

Click Window, Mimic.

Click on house1 in the list of mimics.

Click the Show Labels box and OK.

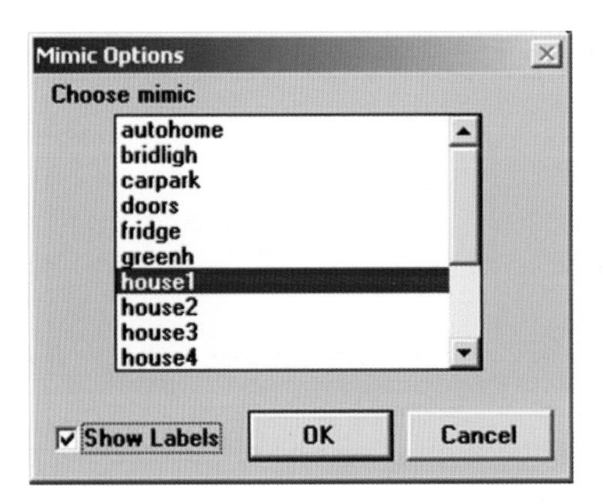

Flowol will load the house1 mimic – your screen will look like the one below.

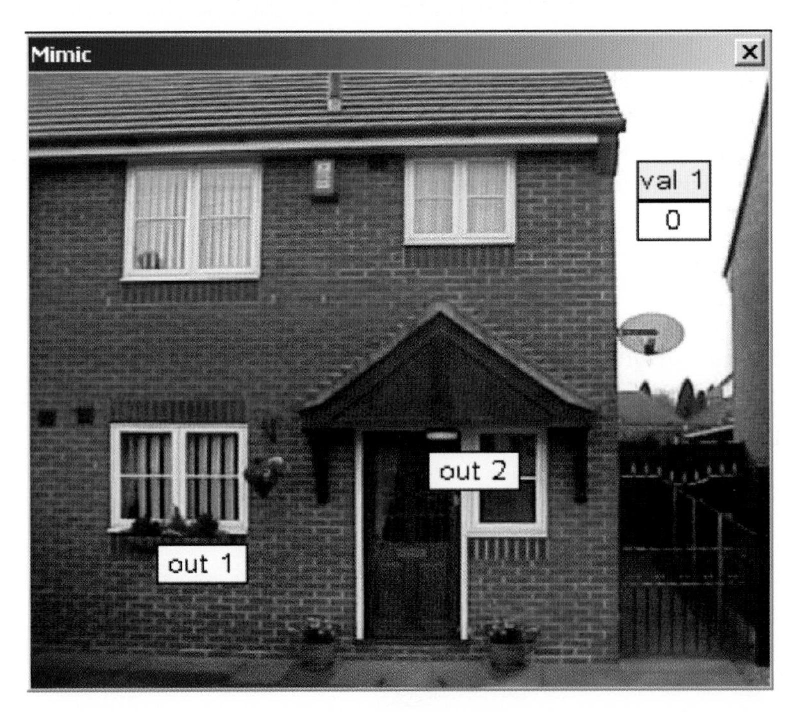

Before the control problem is described, you need to spend a few minutes finding out about the inputs and outputs on this mimic.

This mimic has one input from a light sensor labelled val 1. — val 1

This is an analogue sensor that monitors the light level outside the house. Analogue sensors can send many different values back to a computer rather than just ON or OFF from a digital sensor. The light level detected by this sensor is represented by a value on a scale from 0 to 100. A value of 0 from the sensor means no light is being detected and it is completely dark. A value of 100 from the sensor means it is bright daylight.

Because this is a simulated situation we must control the light level when the mimic is running by changing val 1. We'll try doing this now.

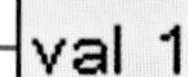

 Rest your mouse pointer inside the box underneath the val 1 label.

 Click a few times on the left mouse button – you will see the value inside the box going up. This simulates the light level detected by the sensor going up as it gets lighter outside the house.

 Click a few times on the right mouse button – you will see the value going down. This simulates the light level detected by the sensor going down as it gets darker outside the house.

This mimic has two outputs labelled out 1 and out 2.

These outputs are connected to security lights inside and outside the house.

 Click in the window just above the out 1 label – which light is this output connected to? — out 1

 Click the white box just above the out 2 label – which light is this output connected to? — out 2

Now you know a little more about this mimic, read on to understand the control problem that needs solving.

The owners of the house want both security lights to come **ON** when it gets dark outside. They want the lights to go **OFF** when it gets light outside. The system must check the light level outside the house all the time and respond straight away whenever it changes.

You can assume it is dark enough for the lights to be turned **ON** when the level of light detected by the light sensor outside is **less than 40**.

- Draw a flowchart to describe this problem on Worksheet 3.
- Use the flowchart you have drawn on Worksheet 3 to help you build a Flowol flowchart to control the house1 mimic.
- Test your flowchart by clicking on val 1 to simulate the light level outside the house changing. Do the right lights come on and go off at the correct times?

If your flowchart doesn't work follow the steps below to check through and correct it.

- Click the Stop button.
- Check through your flowchart making sure that:
  - the right question is being asked inside the DECISION symbol;
  - the YES Arrow is joined to the right outputs;
  - the part of the flowchart leading from the YES Arrow joins back to the top of the flowchart just after the START symbol to complete a loop;
  - the NO Arrow is joined to the right outputs;
  - the part of the flowchart leading from the NO Arrow joins back to the top of the flowchart just after the START symbol to complete a loop.
- When you've finished, label the flowchart with your name and print it.
- Save the flowchart using a suitable file name.
- Close Flowol down.

Chapter 4 Multiple Inputs

In the last chapter you learned how a decision could be used to control loops by checking the input from a sensor. In this chapter you will learn how to solve more complicated control problems where more than one input needs checking.

Make sure you have a copy of Worksheet 4.

The house2 mimic

Load Flowol.

Click Window, Mimic.

Click on house2 in the list of mimics.

Make sure the Show Labels box is checked and click OK.

Flowol will load the house2 mimic.

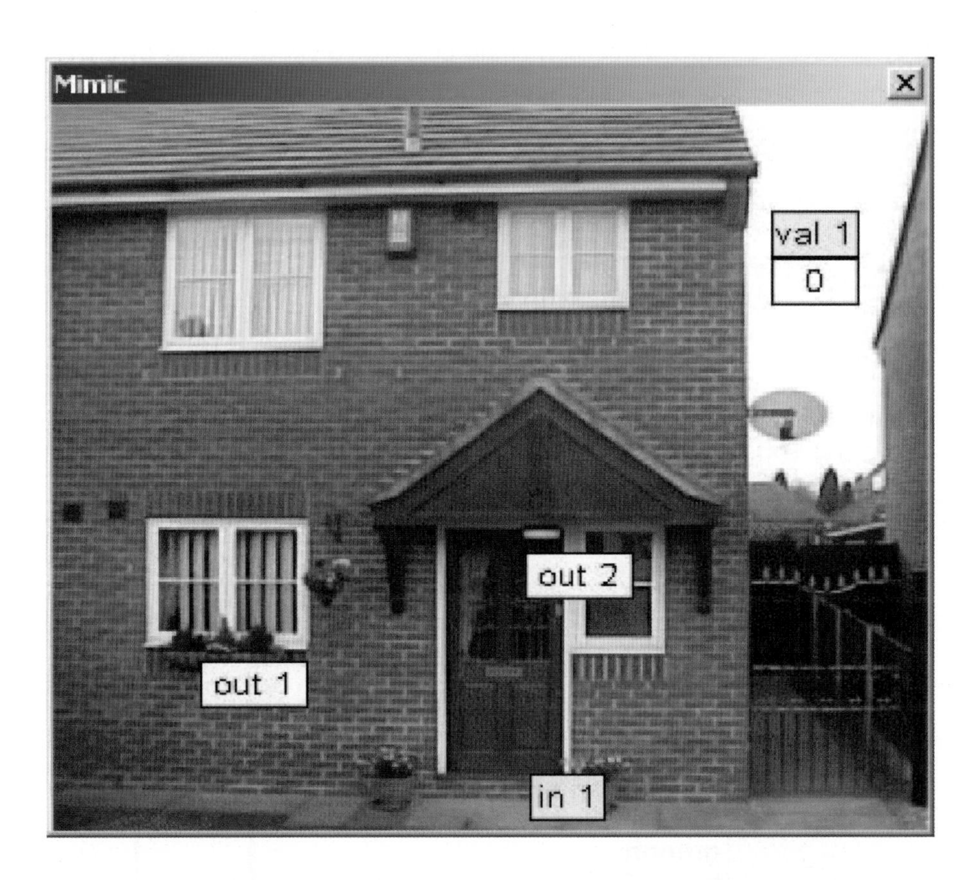

This mimic is very similar to the house1 mimic you used at the end of the last chapter. It has two outputs labelled out 1 and out 2.

These outputs are connected to security lights inside and outside the house. They work in exactly the same way as the ones on the house1 mimic.

The mimic has two inputs.

The sensor labelled val 1 detects the light level outside the house – this works in exactly the same way as the one on the house1 mimic. val 1

There is a second sensor at the front of the house labelled in 1. in 1
This sensor is used to detect movement at the front of the house. We can simulate movement by clicking in 1. We'll try doing this now.

 Rest your mouse pointer on the door of the house.

 Click a few times on the left mouse button – what happens when you do this?

The mimic simulates movement at the front of the house by making a person appear and disappear.

The sensor labelled in 1 will detect this movement when the mimic is running.

in 1

The control problem

Here is a description of the control problem that needs solving.

The owners of the house want the security light **inside** the house to be turned **ON** when it gets dark outside. When this light is on the system must constantly check for movement at the front of the house. If movement is detected, the **outside** light must be turned **ON**. This light must stay on until no movement is detected. When it is light outside, both lights should be turned **OFF**.

You can assume it is dark enough for the lights to be turned on when the level of light detected by the sensor outside is **less than 40**.

The first part of this problem is very similar to the one you tried at the end of the last chapter. The first few steps we need to work through are:

Check the light level outside the house

If the light level is less than 40 turn the light inside the house ON

If the light level is more than 40 turn the light inside the house OFF

The first step will use a DECISION symbol to check the light level outside the house by asking a question about val 1. Depending on the answer to this question OUTPUT symbols will be needed to use to turn the light inside the house either ON or OFF.

val 1

The first half of the flowchart is shown below. This flowchart is only partly completed - some of the symbols are blank and some connecting arrows are missing.

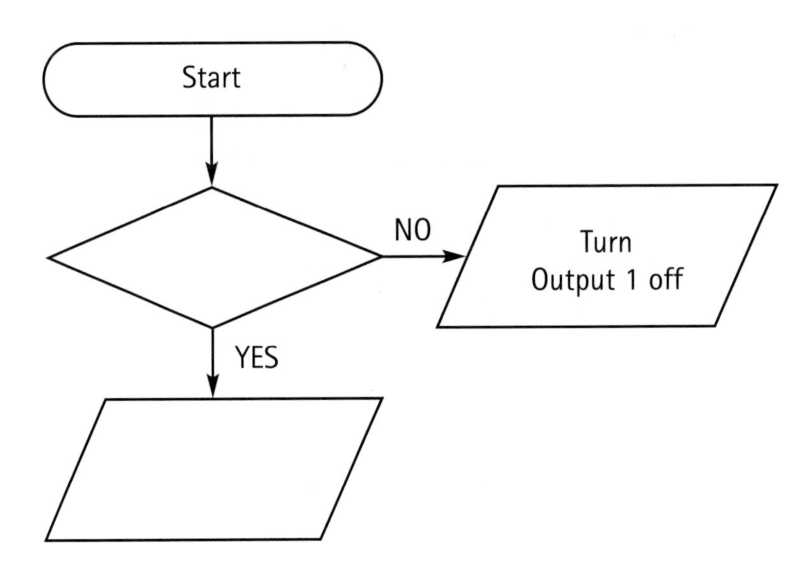

This part of the flowchart is also shown on Worksheet 4. Before starting any work on the computer, you need to plan out the steps to take.

Planning the solution

Complete the blank symbols for this part of the flowchart on your copy of Worksheet 4.
You will need to decide:

- what question must be asked inside the DECISION symbol;
- what the output symbol connected to the YES Arrow must do

The steps we need to go through for the second part of this problem are:

Check for movement outside the house

If movement is detected turn the light outside the house ON

If no movement is detected turn the light outside the house OFF

These steps only need to be carried out when the light inside the house has been turned ON.

We need another DECISION symbol to check for movement outside the house by a sking a question about in 1. Depending on the answer to this question, OUTPUT symbols will be needed to turn the light outside the house either ON or OFF.

in 1

4

Part of the second half of the flowchart is shown below. Some of the symbols are blank and some connecting arrows are missing.

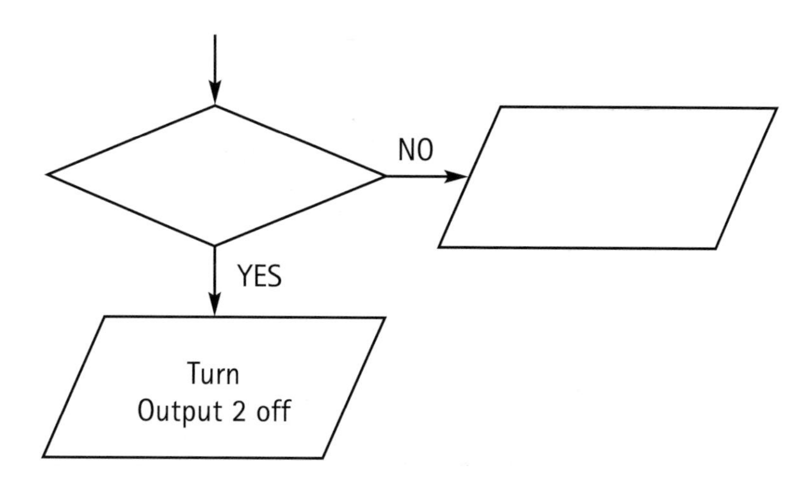

This part of the flowchart is also shown on Worksheet 4.

Complete the blank symbols for this part of the flowchart on your copy of Worksheet 4. You will need to decide:

- what question must be asked inside the DECISION symbol;
- what the output symbol connected to the NO Arrow must do;

Ask your teacher to check your answers on Worksheet 4 before you carry on.

Building the flowchart

Now you are ready to build a flowchart to solve this problem.

Start by loading a blank flowchart like the one you've just completed on Worksheet 4.

Click File, Open... from the main menu at the top of the screen.

The file you need to load is called H2.flo.

Click on the file called H2.flo and click OK.

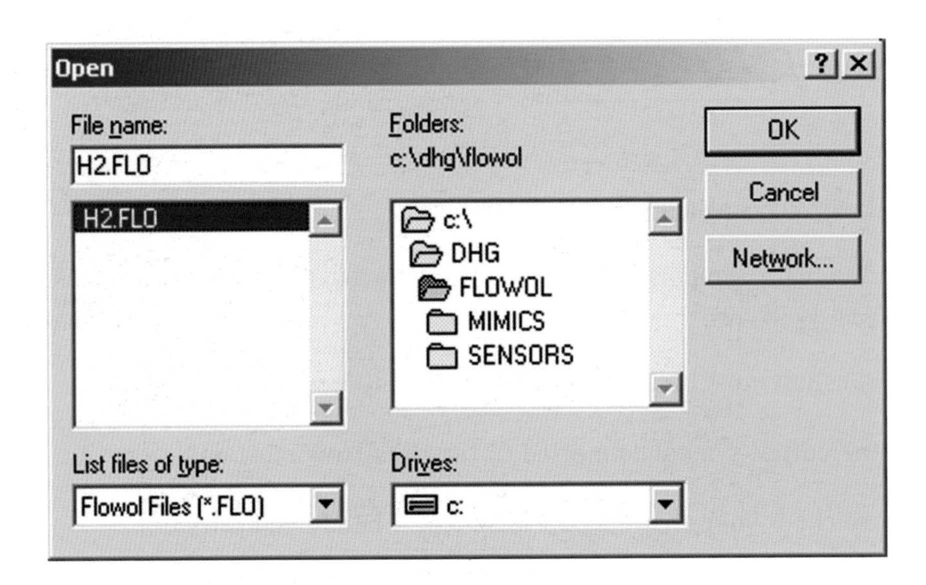

Your teacher will tell you where to find this file if you can't see it in the list on your screen.

If you have not already loaded the house2 mimic, click OK when you see the message shown below.

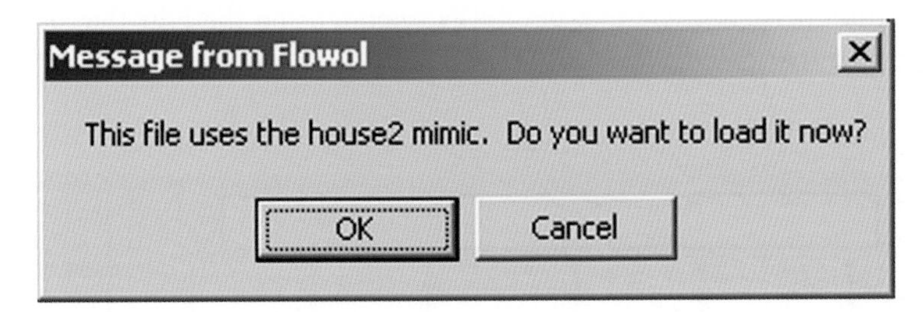

The house2 mimic will be loaded along with a partly completed flowchart like the one on Worksheet 4.

Click Window, Mimic.

Click the Show Labels box and OK.

Your screen should now look like the one below.

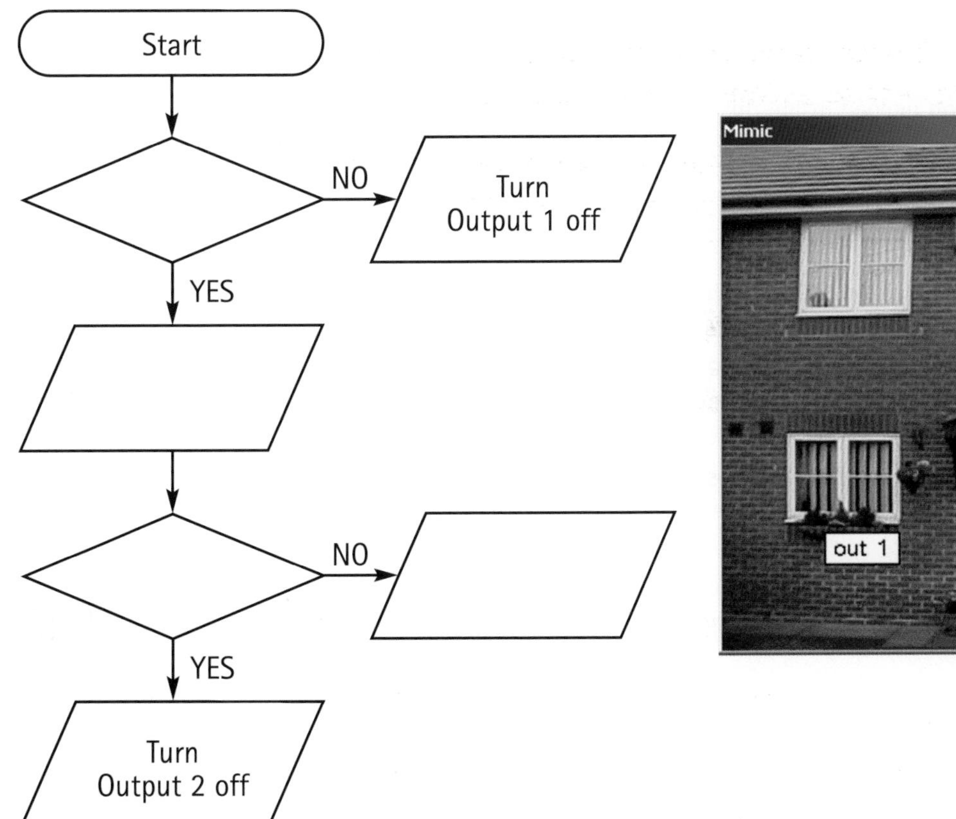

Using your copy of Worksheet 4, follow these instructions to complete the flowchart.

- Start by clicking on the Hand tool, then the first DECISION symbol.
- Build up the question needed inside this DECISION symbol using the Edit Decision box at the bottom of the screen.

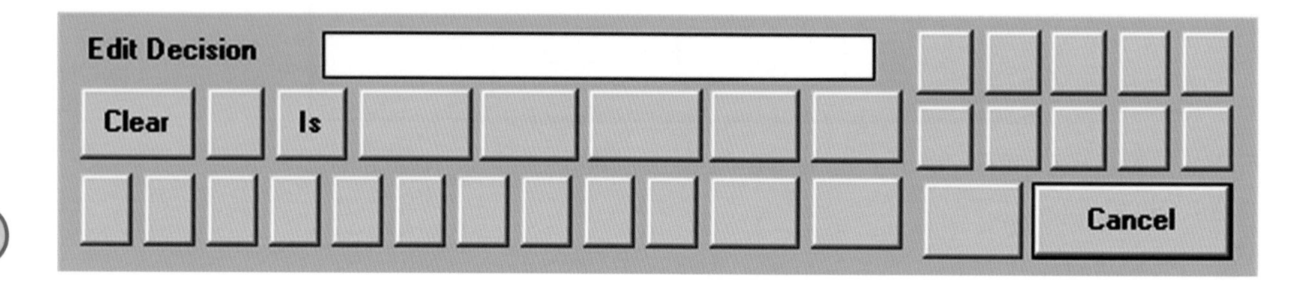

- Click on the Hand tool then the OUTPUT symbol underneath the first DECISION symbol.

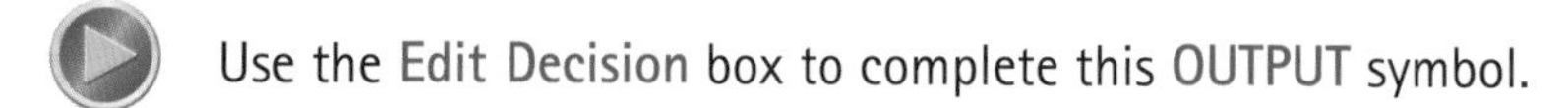

- Use the Edit Decision box to complete this OUTPUT symbol.
- Click on the Hand tool, then the second DECISION symbol.
- Build up the question needed inside this DECISION symbol using the Edit Decision box at the bottom of the screen.
- Click on the hand tool then the OUTPUT symbol on the right of the second DECISION symbol.
- Use the Edit Decision box to complete this OUTPUT symbol.

Next you need to create loops in the flowchart to make the system check continuously if the level of light has changed or if movement has been detected at the front of the house.

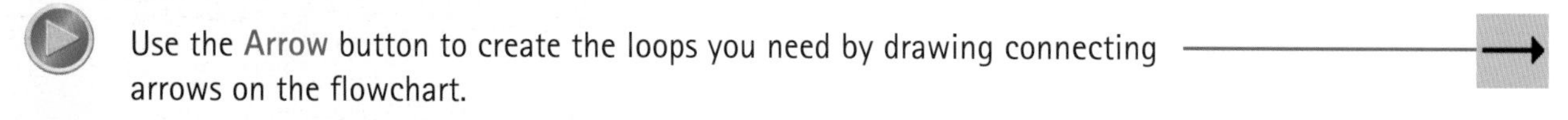

- Use the Arrow button to create the loops you need by drawing connecting arrows on the flowchart.
- If an arrow doesn't go where it should, use the hand tool to select it and the Erase button to remove it before trying again!

ERASE

If you get stuck or don't understand where the connecting arrows need to go to create the loops, ask for help!

Testing the flowchart

Once the symbols have been joined together you can test the flowchart by running it to see what happens.

- Click the **Run** button.
- Test your flowchart by clicking on **val 1** to simulate the light level outside the house changing.

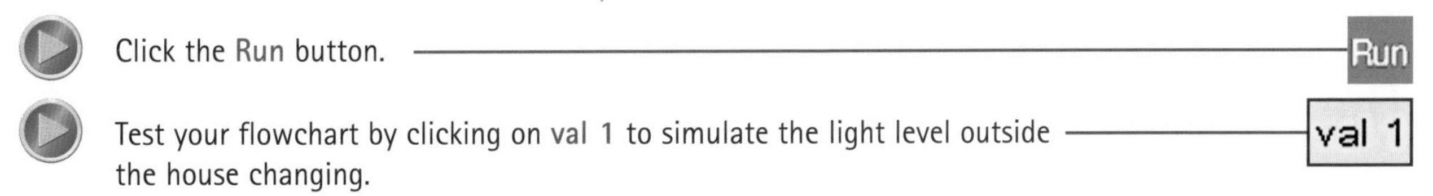

Does the right light come on and go off at the correct times?

- Reduce the light level until the light inside the house comes on.
- Now rest your mouse pointer on the door of the house and click the left mouse button.

Does the right light come on when someone is outside the house?

- Click on the left mouse button again.
- Does the right light go off when there is no one outside the house?
- If your flowchart doesn't work click the **Stop** button, check through and correct any mistakes before running it again.

Stop

If you've checked your flowchart and still can't get it to work, ask for help.

- When you've finished, label the flowchart with your name and print it.
- Save the flowchart using a suitable file name.
- Close Flowol down.

Chapter 5 Controlling a Pedestrian Crossing

In the last chapter you leaned how to solve more complicated control problems where more than one input needed checking. In this chapter you'll use these skills to control the lights at a pelican crossing.

Make sure you have a copy of Worksheet 5.

The problem

To solve this problem you need to think about the things that happen at a pelican crossing when someone crosses the road. The green lights and red man stay on until the Cross button is pressed – then what happens? You need to work out exactly what must happen before you can build a Flowol flowchart to control the pelican crossing mimic.

Worksheet 5 lists the things that happen after the Cross button is pressed – but some words are missing, ready for you to fill them in.

To help you complete Worksheet 5 you are going to watch a PowerPoint presentation called crossing1.ppt. Your teacher will tell you how to load and run this file.

 Watch the crossing1.ppt presentation.

 Complete the missing words on Worksheet 5.

This presentation will run over and over again until you press the Esc key.

 When you have finished, press the Esc key and close PowerPoint.

Before carrying on, check the steps you have written down on Worksheet 5 with your teacher.

Now that you have broken down the problem into smaller steps, you can draw a flowchart.

Draw a flowchart on Worksheet 5 to describe the problem – remember this is another problem that has no end; it just keeps on repeating forever.

Before you carry on ask your teacher to check your flowchart.

Getting started

Now you can get started building a Flowol flowchart to control this pedestrian crossing – follow the steps below to do this.

Load Flowol.

You need to load a mimic called pelican.

Click Window, Mimic.

Click on pelican in the list of mimics.

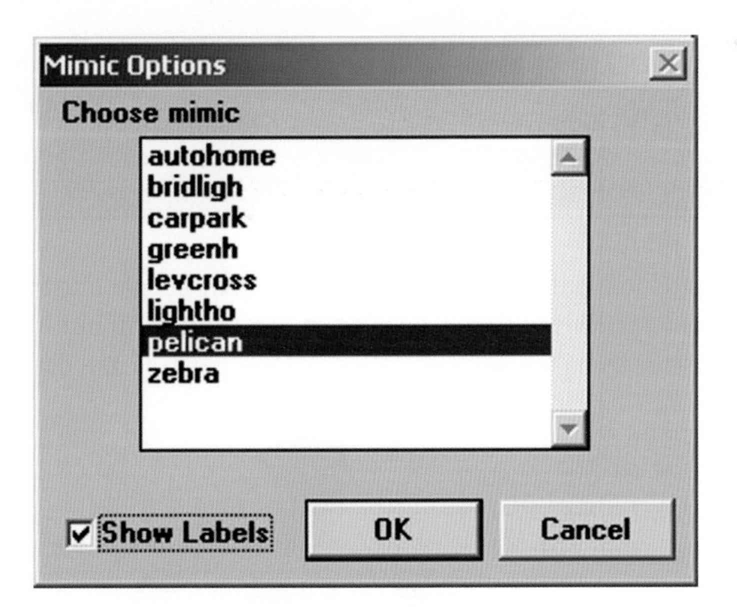

Click the Show Labels box.

Click OK.

Flowol will load the pelican mimic.

This mimic has one input. The input comes from the button that people must press to make the traffic lights change. This button is labelled in 1 – short for input 1 – on the mimic. When the mimic is running we will simulate someone pressing the button by clicking on this input.

in 1

This mimic has six outputs. Before we start building a flowchart to control the crossing we must find out what each output does.

Look at your copy of Worksheet 5 – you will see a table like the one below.

Green lights on	TURN OUTPUT	ON
Green lights off	TURN OUTPUT	OFF
Amber lights on	TURN OUTPUT	ON
Amber lights off	TURN OUTPUT	OFF
Red lights on	TURN OUTPUT	ON
Red lights off	TURN OUTPUT	OFF
Green man on	TURN OUTPUT	ON
Green man off	TURN OUTPUT	OFF
Red man on	TURN OUTPUT	ON
Red man off	TURN OUTPUT	OFF
Wait sign on	TURN OUTPUT	ON
Wait sign off	TURN OUTPUT	OFF

You are going to use the pelican mimic to complete this table to show what the different outputs do when they are turned ON and OFF. This will be useful when you start to build the Flowol flowchart to control the crossing.

To get you started, here's how to complete the first row of the table.

Click on the light next to the label out 1 on the mimic – what happens? The green lights should come on. So, to turn the green lights on we must TURN OUTPUT 1 ON. Write 1 in the empty space on the first row of your table so that it looks like the one below.

out 1

Green lights on	TURN OUTPUT	1	ON
Green lights off	TURN OUTPUT		OFF

Now use the mimic to complete the rest of the table on Worksheet 5.

Building the flowchart

Put a START symbol at the top of the screen.

We've already said that the green lights and red man must stay ON until the Cross button is pressed – we must use an OUTPUT symbol do this.

Put an OUTPUT symbol underneath the START symbol.

This output symbol must send a signal to turn both the green lights and red man ON. Look at your copy of Worksheet 5 – you'll see we need to turn out 1 and out 8 ON for this to happen.

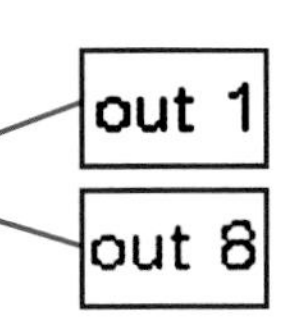

Use the Edit Output box at the bottom of the screen to make this OUTPUT symbol turn out 1 and out 8 ON.

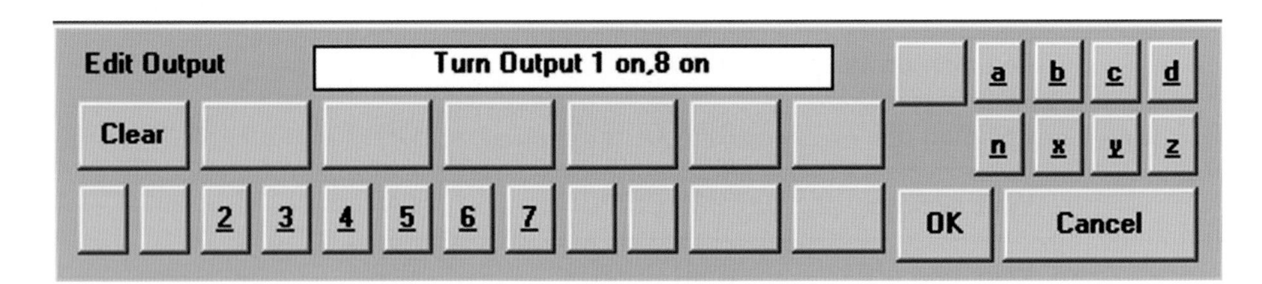

Your flowchart should now look like the one below.

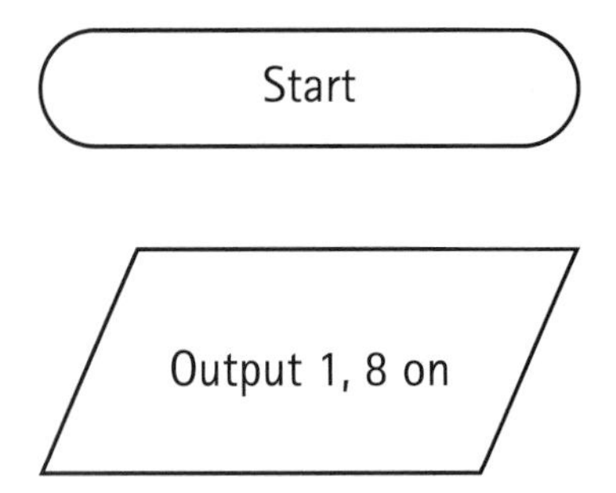

The green lights and red man must stay ON until the Cross button is pressed. The next symbol in the flowchart must ask a question to check if the Cross button has been pressed. To ask a question we'll need to use a DECISION symbol.

 Put a DECISION symbol just underneath the OUTPUT symbol.

This DECISION symbol needs to ask the question, Is input 1 on?

 Build up this question inside the DECISION symbol using the Edit Decision box at the bottom of the screen.

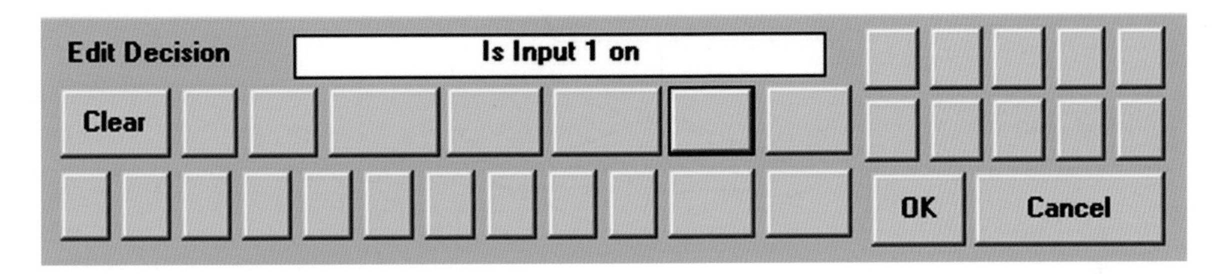

Your flowchart should now look like this:

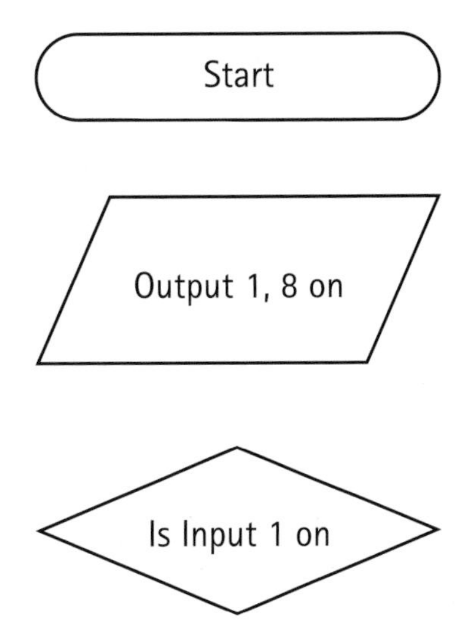

The next thing we need to do is add an arrow to the flowchart to show what must happen when the answer to the question in the DECISION symbol is NO. This will be the case when the cross button hasn't been pressed, so in 1 will be OFF.

in 1

We'll join the first three symbols on the flowchart together.

 Use the Arrow button to join the first three symbols together so that your flowchart looks like the one on the next page.

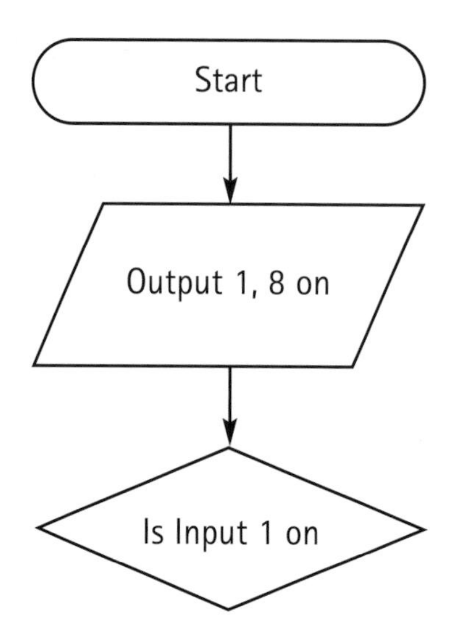

We need to create a loop joining the DECISION symbol back to the top of the flowchart with a NO Arrow. This will keep the green lights and red man ON until the Cross button is pressed, making the answer to the question inside the DECISION symbol YES.

Use the NO Arrow button to join the DECISION symbol back to the top of the flowchart. NO →

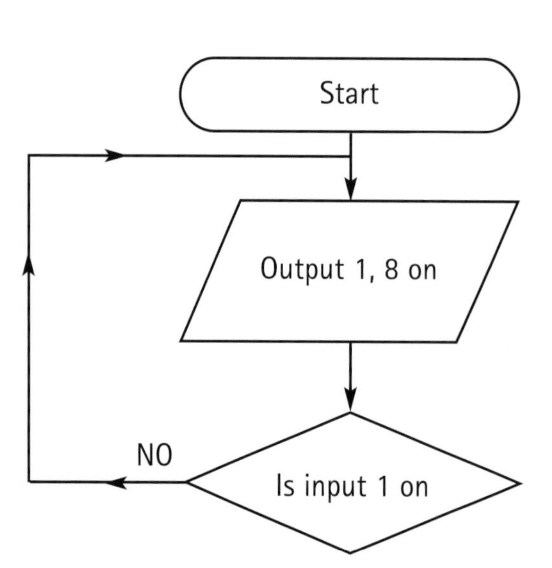

Next we need to build the rest of the flowchart to show what must happen if the answer to the question inside the DECISION symbol is YES. These are the steps that you worked out earlier on Worksheet 5. The first step was to turn the Wait sign ON. We'll work through this step before you finish the rest of the flowchart on your own.

To turn the Wait sign ON you need another OUTPUT symbol.

Put an OUTPUT symbol underneath the DECISION symbol on your flowchart.

This OUTPUT symbol must tell Flowol to send a signal to turn the Wait sign ON. If you look at your table on Worksheet 5 you will see you need to Turn Output 5 on for this to happen.

Use the Edit Output box at the bottom of the screen to set this OUTPUT symbol to Turn Output 5 on.

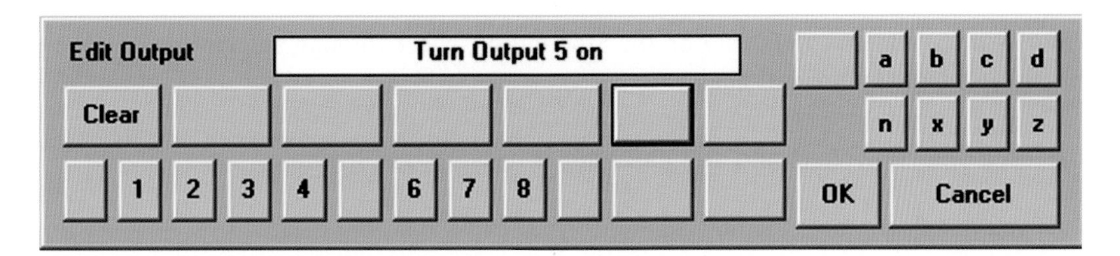

The next thing we need to do is add an arrow to the flowchart to show that the instructions in this OUTPUT symbol must be carried out if the answer to the question inside the DECISION symbol is YES.

Use the YES Arrow button to join the DECISION symbol to the OUTPUT symbol underneath it. 

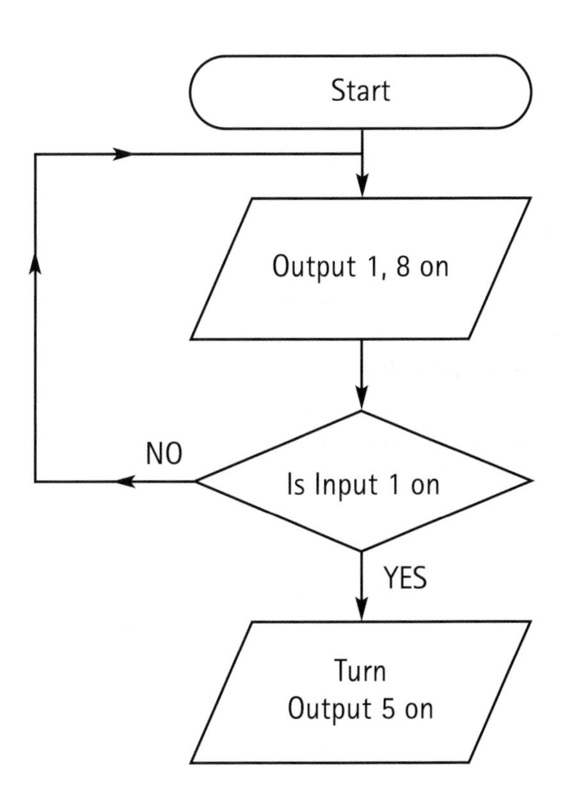

Now complete the rest of the flowchart yourself. Use the table on Worksheet 5 and your own flowchart drawing to help you.

If you haven't completed Worksheet 5 ask your teacher for a copy of the crossing1 help sheet.

 When you've finished click the **Run** button on the left of the screen. — Run

Flowol will run the flowchart – click on the **Cross** button on the mimic. The lights should change just like those on the crossing you watched in the presentation at the beginning of this chapter.

 If your flowchart doesn't work click on the **Stop** button, check through and correct any mistakes before running it again. — Stop

If you've checked your flowchart and still can't get it to work, ask for help.

 When you've finished label the flowchart with your name and print it out.

 Save the flowchart using a suitable file name.

 Close Flowol down.

Chapter 6 Subroutines

In the last chapter you practised what you had learnt so far to create a flowchart that controlled a pedestrian crossing. In this chapter you will learn how to make systems more efficient by using subroutines.

Getting started

To get started you are going to watch a PowerPoint presentation called crossing2.ppt. Your teacher will tell you how to load this file.

Watch the crossing2.ppt presentation.

What is different about this pedestrian crossing compared to the one in the presentation you watched at the beginning of the last chapter?

The presentation will run over and over again until you press the Esc key.

When you have finished press the Esc key and close PowerPoint down.

Did you notice anything different about this crossing? If you thought the way the lights changed was different you're right. At this crossing, the red light is turned off, and then the amber lights and green man flash a number of times before the green lights and red man are turned back on.

The flowchart you built in the last chapter didn't do this. To make the amber lights and green man flash you will need to add some extra symbols to the flowchart.

The pelican crossing mimic is shown below.

The amber lights and green man are labelled out 2 and out 7 on this mimic.

out 2

out 7

6 The extra symbols needed to make the amber lights and the green man flash once are shown below.

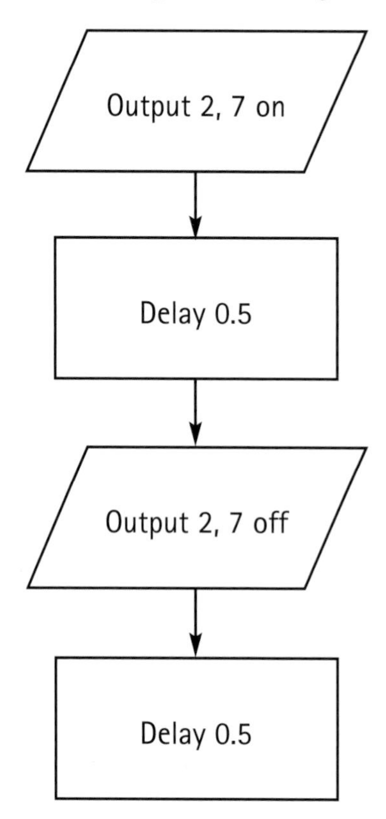

These symbols make the amber lights and green man flash by:

turning out 2 and out 7 ON

waiting for half a second

turning out 2 and out 7 OFF

Suppose we wanted the amber lights and green man to flash five times. How many extra symbols would we have to use? Let's load a flowchart that does just this and find out!

Load Flowol.

Click File, Open... from the main menu at the top of the screen.

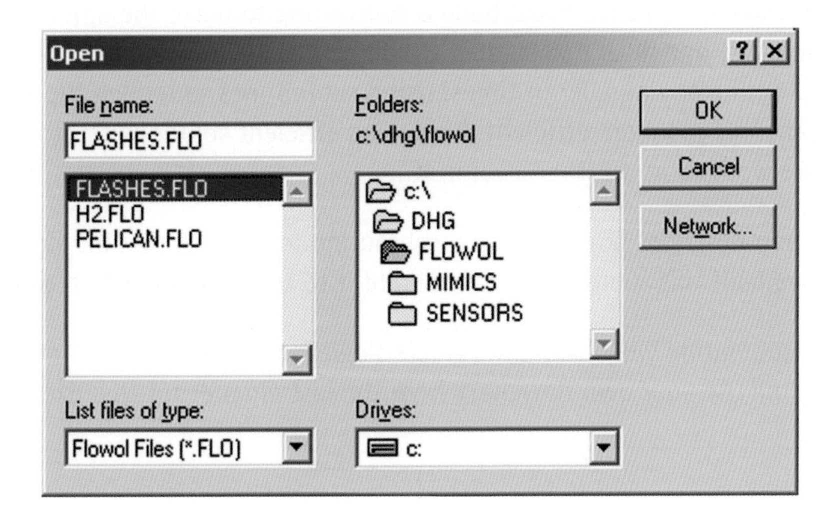

The file you need to load is called flashes.flo.

Click on the file called flashes.flo and click Open. Your teacher will tell you where to find this file if you can't see it in the list on your screen.

Click OK when you will see the message shown below.

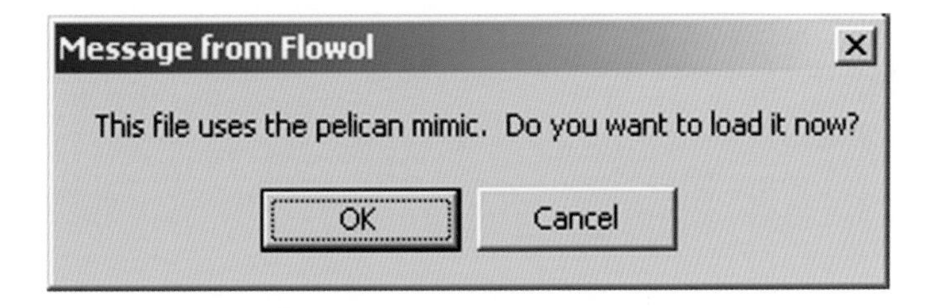

The pelican mimic will be loaded along with a completed flowchart.

Click the Run button. —— Run

Count the number of times the amber lights and green man flash - you can run the flowchart again if you need to.

Use the scroll bar on the right of the screen to work your way down through the flowchart and count the number of symbols in between the START and STOP symbols.

How many symbols did you count?

This is the number of extra symbols you would need to add to your flowchart from the last chapter to make the amber lights and green man flash just five times!

Instead of using all these symbols we could build a subroutine to make the amber lights and green man flash once. Then, just one PROCESS symbol would be needed in the main flowchart to repeat the steps in the subroutine any number of times. Using subroutines to reduce the number of steps in a flowchart makes them much more efficient. Producing efficient solutions to problems is one way to show that you are working at a higher level in ICT.

The rest of this chapter describes how to build a subroutine to make the amber lights and green man flash. When you have built this subroutine you will add it to the main flowchart and test it.

Before building the subroutine we'll load an example flowchart solution for the pedestrian crossing. Alternatively, you could use your own flowchart from the last chapter for this exercise if you are sure it is correct.

Load Flowol.

Click File, Open... from the main menu at the top of the screen.

If you see the message below click No.

The file you need to load is called pelican.flo.

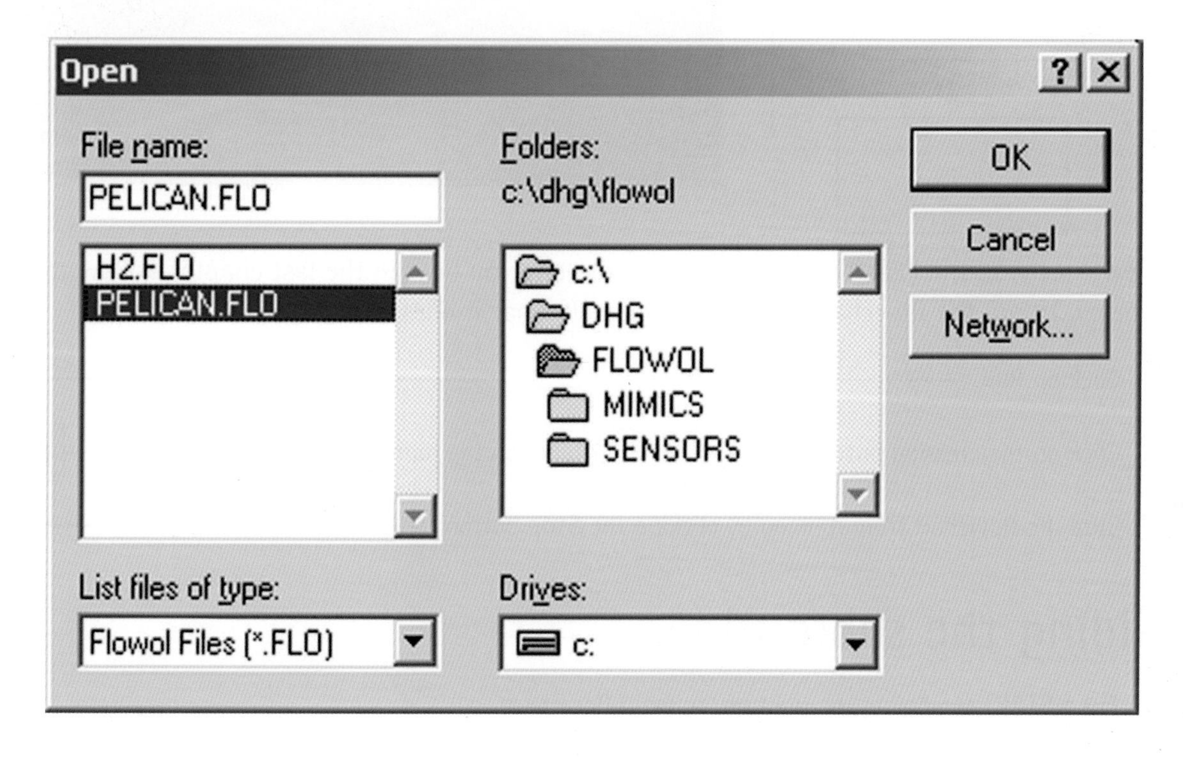

Click on the file called pelican.flo and click OK.
Your teacher will tell you where to find this file if you can't see it in the list on your screen.

The pelican mimic will be loaded along with a flowchart. Not all the flowchart is shown in the picture below. You can use the scroll bar on the right of your screen to move down and look at the rest of the flowchart if you want to.

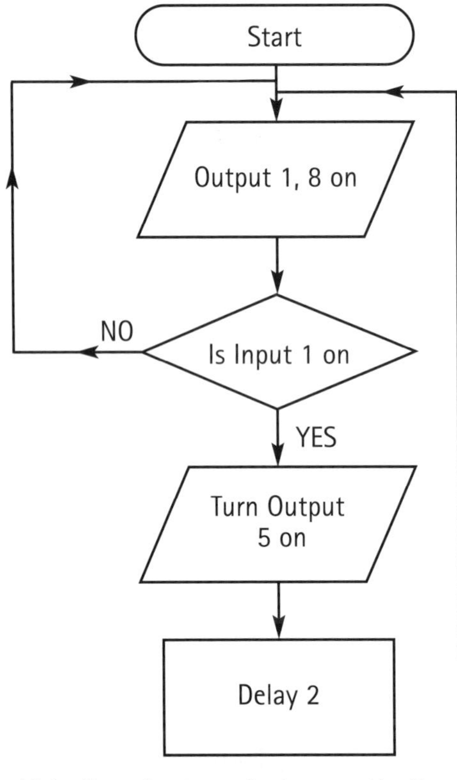

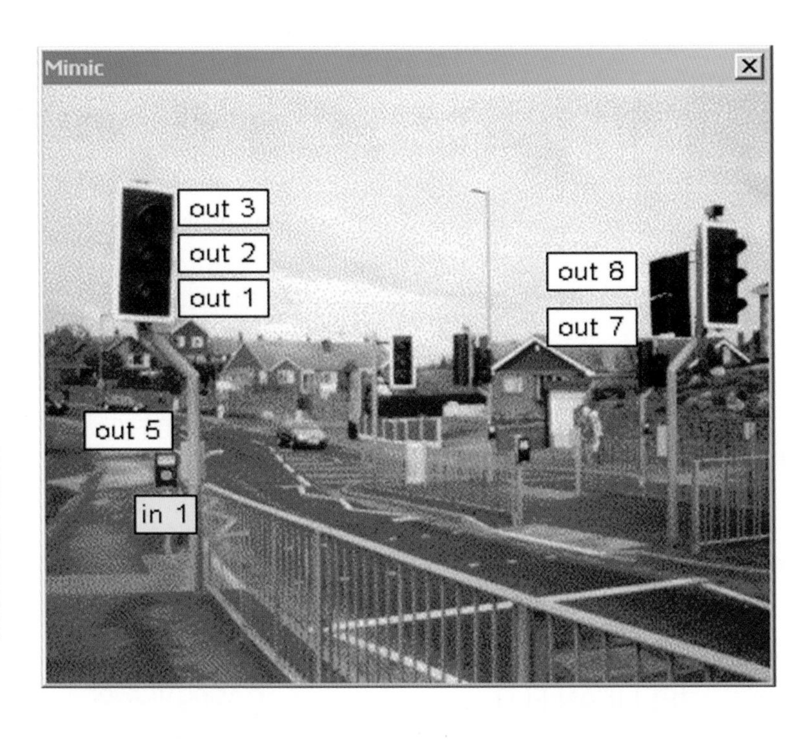

6

This flowchart works in exactly the same way as the one you built in the last chapter. Follow the steps below to run the flowchart and remind yourself what it does.

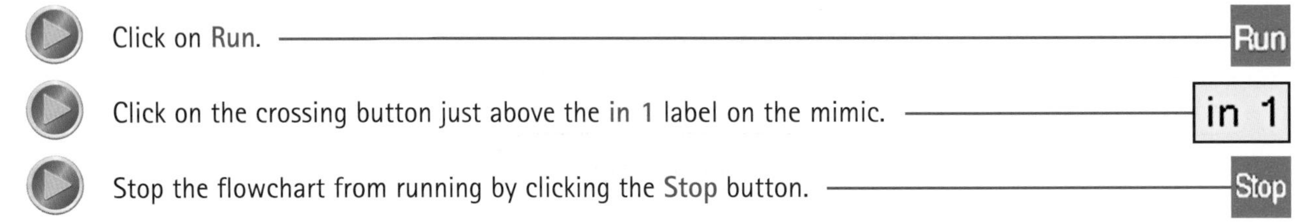

- Click on Run.
- Click on the crossing button just above the in 1 label on the mimic.
- Stop the flowchart from running by clicking the Stop button.

Building the subroutine

- Put a START symbol on the right of the main flowchart.

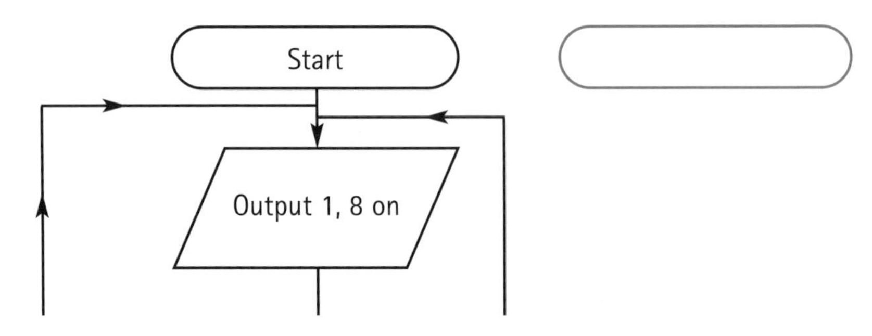

- Click the Sub button and type flash – the name for this subroutine – then click OK.

Sub

OK

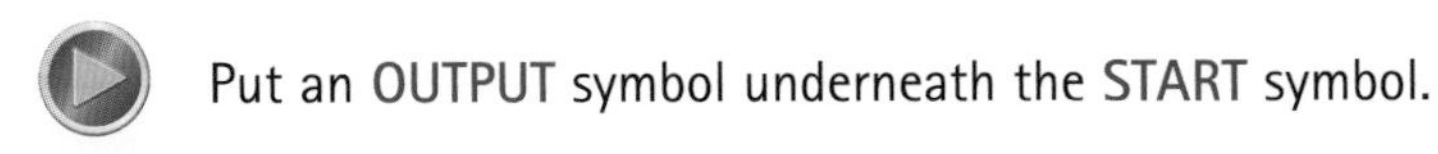

- Put an OUTPUT symbol underneath the START symbol.
- Set this output symbol to turn output 2 and 7 ON using the Edit Output box at the bottom of the screen.
- Use the Arrow button to join these two symbols together.

Your subroutine flowchart should look like the one below.

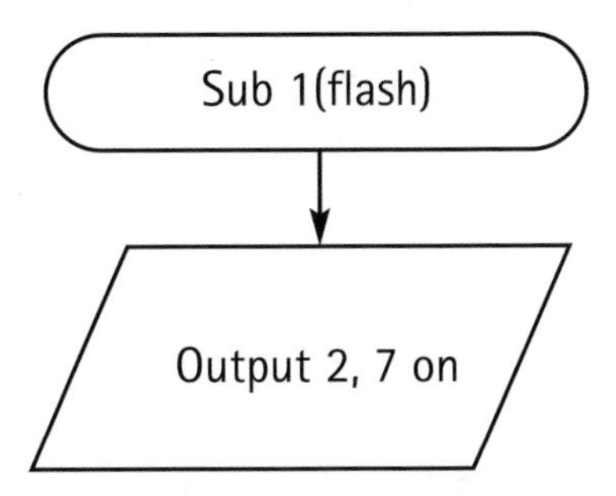

- Put a PROCESS symbol underneath the OUTPUT symbol.
- Set this PROCESS symbol to give a 0.5 second delay using the Edit Process box at the bottom of the screen.
- Use the Arrow button to join these two symbols together.

Your subroutine flowchart should look like the one below.

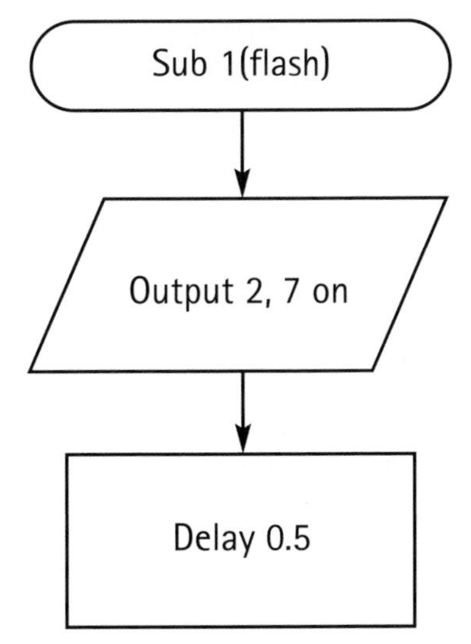

- Put another OUTPUT symbol underneath the Delay 0.5 symbol.
- Set this output symbol to turn output 2 and 7 OFF using the Edit Output box at the bottom of the screen.

Use the Arrow button to join this symbol to the Delay symbol above. →

Your subroutine flowchart should look like the one below.

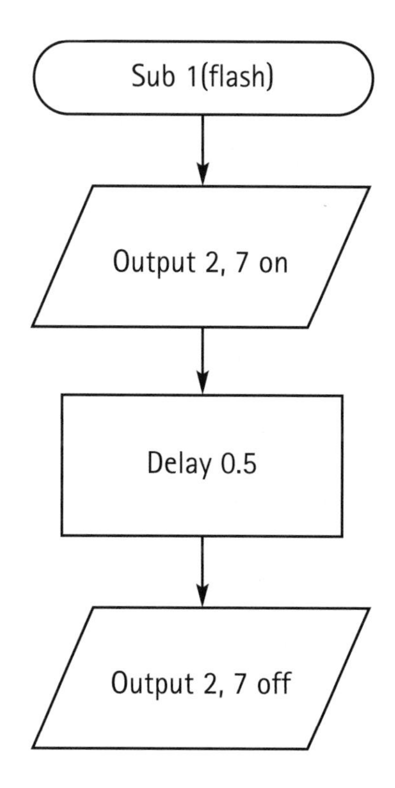

Put another PROCESS symbol underneath the OUTPUT symbol.

Set this process symbol to give another 0.5 second delay using the Edit Process box at the bottom of the screen.

Put a STOP symbol at the bottom of the subroutine.

Use the Arrow button to join these symbols together. →

The flowchart for the subroutine is finished and should look like the one below.

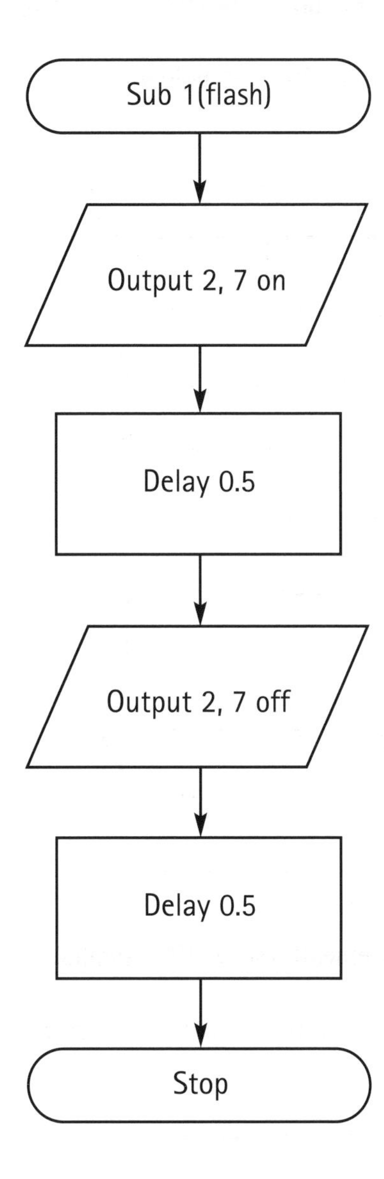

That's the subroutine finished - now we need to add it to the main flowchart and test it.

6

Using a subroutine in a flowchart

The part of the main flowchart where the new subroutine must go is shown below.

Use the scroll bar on the right of the screen to move down until you can see this part of the flowchart.

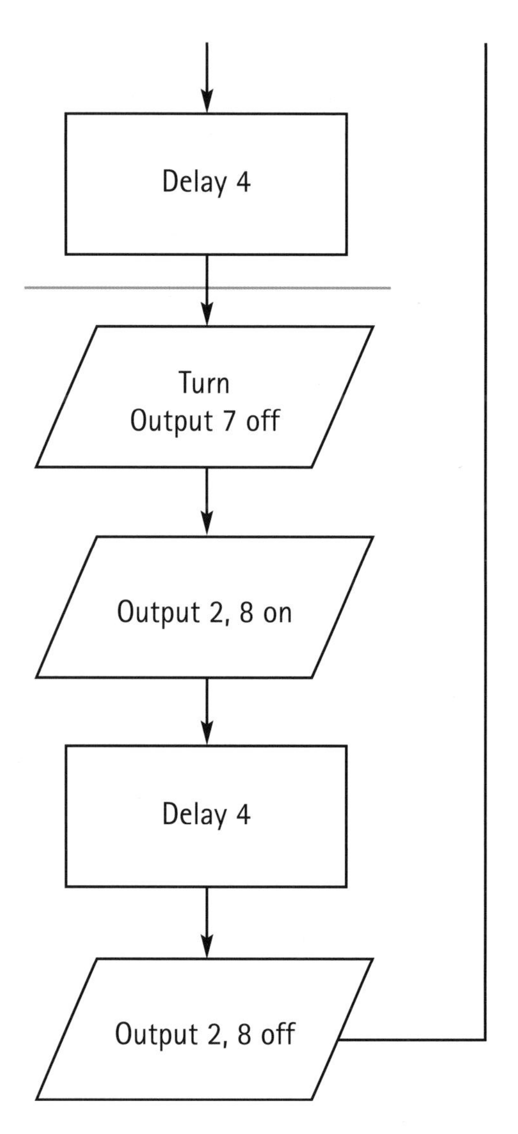

Use the Hand tool and Erase button to remove all the symbols underneath the green line – the new subroutine will replace these symbols.

Before we use the new subroutine we need to turn the red traffic light OFF.

- Put an OUTPUT symbol underneath the Delay symbol at the bottom of the flowchart.
- Use the Edit Output box at the bottom of the screen to make this OUTPUT symbol turn OFF the red traffic light, labelled out 3 on the mimic. — out 3
- Use the Arrow button to join this symbol to the rest of the flowchart. —→

Once the red traffic light is turned OFF the amber traffic lights and green man must start to flash. Follow the steps below to use the subroutine in the flowchart.

- Put a PROCESS symbol underneath the OUTPUT symbol at the bottom of the flowchart.
- Click on Sub in the Edit Process box at the bottom of the screen. — Sub
- Click on 1(flash). — 1(flash)
- Click on 5. This will make the subroutine repeat five times. — 5

The Edit Process box on your screen should now look like the one below.

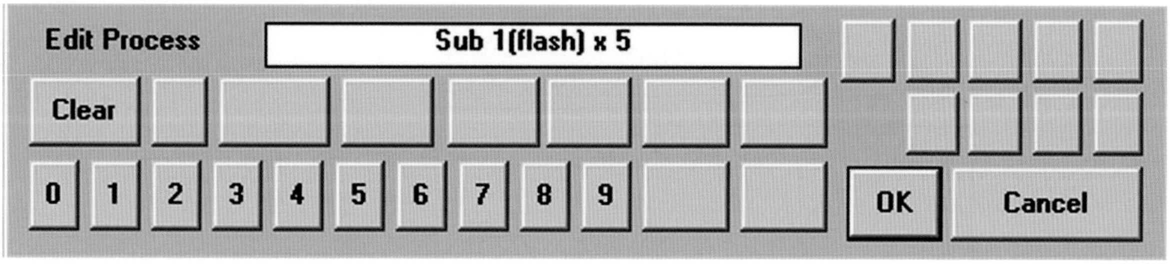

Edit Process	Sub 1(flash) x 5
Clear	
0 1 2 3 4 5 6 7 8 9	OK Cancel

- If the Edit Process box doesn't look like this just click on Clear and repeat the steps above. — Clear
- When you have finished click on OK. — OK
- Use the Arrow button to join this symbol to the rest of the flowchart. —→

The bottom part of your flowchart should now look like the one below.

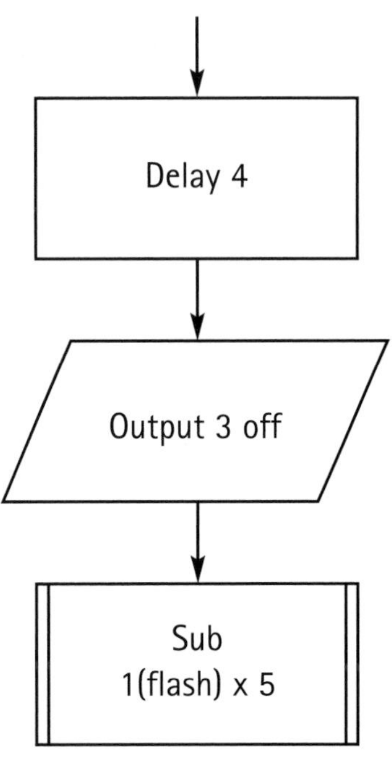

6 Use the Arrow button to join this symbol back to the top of the flowchart.

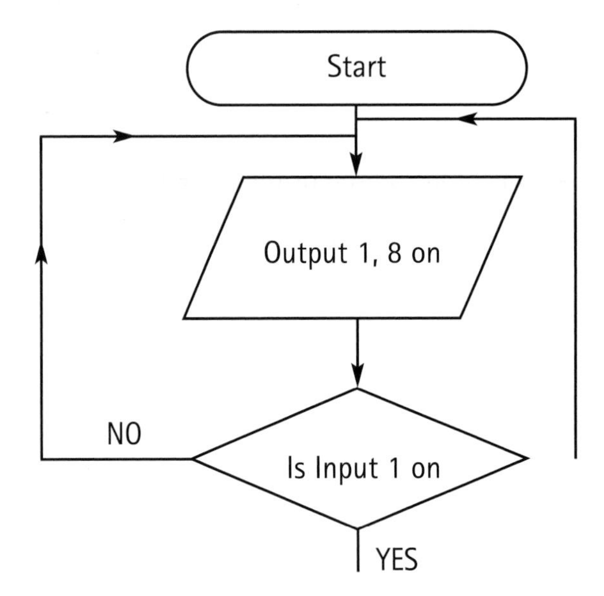

Follow the steps below to run and test the flowchart.

 Click the Run button. 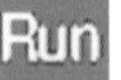

Click on the crossing button just above the in 1 label on the mimic. in 1

Does this crossing work like the one you watched in the presentation at the start of this chapter?

If your flowchart doesn't work follow the steps below to check through and correct any mistakes.

Check that:

- your subroutine is the same as the one on page 66;
- the last two symbols are the same as those shown on page 68;
- the last symbol is joined back to the top of the flowchart just after the START symbol to complete a loop.

When you've finished, label the flowchart with your name and print it out.

Save the flowchart using a suitable file name.

Task C

This task will help you become more confident in building and using subroutines to make systems more efficient.

To get started you will need to load another example flowchart.

Click File, Open... from the main menu at the top of the screen.

If you see the message asking if you wish to save, click No.

The file you need to load is called lightho1.flo.

Click on the file called lightho1.flo and Open.

Your teacher will tell you where to find this file if you can't see it in the list on your screen.

This flowchart uses a mimic called lightho – if you see the message below click OK.

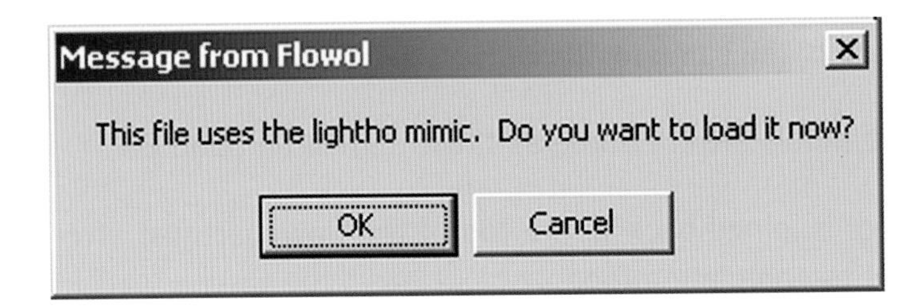

The lightho mimic will be loaded along with a flowchart.

Click the Run button.

Run

The lighthouse light will start flashing. Watch the light as it flashes - does it just flash ON and OFF or is there a special pattern?

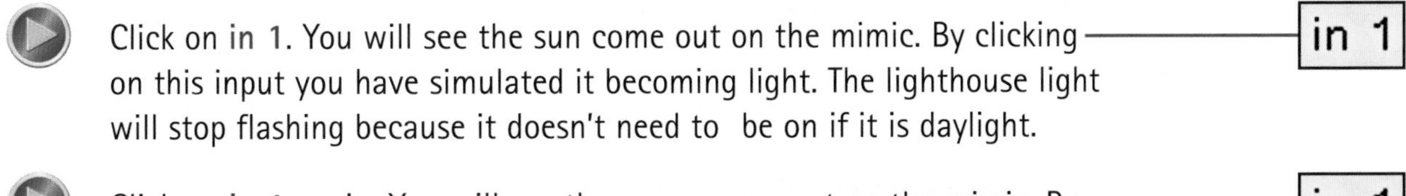

Click on in 1. You will see the sun come out on the mimic. By clicking on this input you have simulated it becoming light. The lighthouse light will stop flashing because it doesn't need to be on if it is daylight.

Click on in 1 again. You will see the moon come out on the mimic. By clicking on this input again you have simulated it becoming dark. The lighthouse light will start flashing because it must be on if it is dark outside.

in 1

Watch carefully – you will see there are two short flashes followed by three long flashes. This pattern of long and short flashes is used to identify the lighthouse. Real lighthouses each have their own special flashing light pattern so that ships at sea can identify them.

Click the Stop button.

The problem with this flowchart is that it is not efficient, because many of the same symbols are repeated. This problem is similar to the flashing amber lights and green man at the pedestrian crossing. We can make this flowchart more efficient by building subroutines and using them to replace the repeating groups of symbols.

Use the scroll bar to work your way down through the flowchart and think about these questions:

- which group of symbols makes the lights flash quickly?
- how many times does this group of symbols appear?
- which group of symbols makes the lights flash slowly?
- how many times does this group of symbols appear?

Your task is to build a new more efficient flowchart to make the lighthouse light flash in exactly the same way. You should do this by building two subroutines and using them in the main flowchart. One subroutine should make the lighthouse light flash once quickly. One subroutine should make the lighthouse light flash once slowly.

- To get started click File, New from the main menu at the top of the screen.
- If you see the message asking if you wish to save, click No.
- When you've finished, test your flowchart.
- If your flowchart doesn't work click on the Stop button, check through and correct any mistakes before running it again. — Stop

If you've checked your flowchart and still can't get it to work ask for help!

- When you've finished, label the flowchart with your name and print it out.
- Save the flowchart using a suitable file name.
- Close Flowol down.

Assessing Yourself

Name: ______________________ Class: ______________

Teacher: ______________________ Date: ______________

Think about the work you have done as you have worked through this book. Put a tick next to the sentences below that describe what you have done.

Level 3

I have built a flowchart to control a simple sequence of events using a Flowol mimic. (Chapters 1 – 2 and Task A) ☐

Level 4

I have planned and built a flowchart to control events using a Flowol mimic. (Chapter 3 and Task B) ☐

I have investigated the effect of changing inputs to test a flowchart and make sure it works correctly. (Chapter 4) ☐

Level 5

I have used decision symbols to control loops in flowcharts by checking input from sensors. (Chapters 5 – 6) ☐

I have improved a flowchart to make it more efficient by replacing repeating groups of symbols with subroutines. (Task C) ☐

Index

Basic ICT Skills series

The books in the Basic ICT Skills series are designed to teach skills in different software packages to pupils from Primary School age upwards. Most of the Basic books have an accompanying 64-page teacher's book. These contain tips for each chapter and photocopiable worksheets which can be completed away from the computer.

For the Further books and some Basic books extra downloadable assignments are available.

Ordering information

Order on-line at
www.payne-gallway.co.uk
E-mail: orders@payne-gallway.co.uk

Other publications:

Title	Publication date
Basic Word 2000-2003	Sep 2004
Basic PowerPoint 2002	May 2003
Basic Internet (3rd Edition)	Aug 2002
Basic Publisher 2000	Aug 2000
Basic Publisher 2002	Aug 2002
Basic Windows 2000/Me	Oct 2000
Basic Web Page Creation using Word 2000-2002	Jun 2003
Basic Access 2000-2003	Sep 2004
Basic Web Pages using Publisher 2002	Nov 2001
Basic HTML	Jan 2002
Basic Paint Shop Pro 8	Jan 2004
Basic Control with Logo	May 2003
Further Word 2000-2002	Apr 2003
Further Excel 2000-2003	Sep 2004
Further Access 2000-2003	Sep 2004
Web Publishing with FrontPage Express	June 2004

Our web site at www.payne-gallway.co.uk gives full details of latest titles and prices.